全国船舶工业职业教育教学指导委员会"十三五"规划教材

船舶防腐与涂装检验专业英语

主　编　曹　雪
主　审　王　宏

哈尔滨工程大学出版社
Harbin Engineering University Press

内容简介

本教材概括性地介绍了船舶涂装及相关专业知识。内容主要涉及金属腐蚀、船舶涂料、船舶涂装步骤、表面处理、涂装环境监测、涂装方法和涂层缺陷等。共分5个项目，18个任务。

本教材可作为船舶涂装工程类专业三年制高职或五年制高职的全日制教材，也可供造船企业、船舶维修及船务公司等相关人员参考使用。

图书在版编目（CIP）数据

船舶防腐与涂装检验专业英语 / 曹雪主编． -- 哈尔滨：哈尔滨工程大学出版社，2022.2
 ISBN 978-7-5661-3396-0

Ⅰ．①船… Ⅱ．①曹… Ⅲ．①船舶－防腐－英语②船体涂漆－英语 Ⅳ．① U672.7 ② U671.91

中国版本图书馆 CIP 数据核字（2022）第 013685 号

船舶防腐与涂装检验专业英语
CHUANBO FANGFU YU TUZHUANG JIANYAN ZHUANYE YINGYU

选题策划	史大伟　薛　力
责任编辑	张　彦　曹篮心
封面设计	李海波

出版发行	哈尔滨工程大学出版社
社　　址	哈尔滨市南岗区南通大街 145 号
邮政编码	150001
发行电话	0451-82519328
传　　真	0451-82519699
经　　销	新华书店
印　　刷	哈尔滨市石桥印务有限公司
开　　本	787 mm×1 092 mm　1/16
印　　张	7
字　　数	235 千字
版　　次	2022 年 2 月第 1 版
印　　次	2022 年 2 月第 1 次印刷
定　　价	28.00 元

http://www.hrbeupress.com
E-mail:heupress@hrbeu.edu.cn

前言 Preface

根据全国船舶工业职业教育教学指导委员会"十三五"规划，以船舶行业的发展方向和船舶涂装工程技术的前沿趋势为引领，依据渤海船舶职业学院"双高"建设项目方案总体目标和要求，结合全国船舶行业指导委员会船舶涂装工程技术专业教学标准，实施《船舶防腐与涂装检验专业英语》教材建设工作。

以充分吸取以往涂装专业教材精髓为出发点，本教材在编写过程中阅读了大量国内外船舶涂装前沿资料，务求做到内容实用、与时俱进。教学材料选择上，以目前船舶制造行业中的涂装设计员、质检员等岗位群为出发点；内容上，以涂装施工管理、材料工时定额等技术领域为基础。改变以往大多数教材只注重船舶原理和结构方面的阐述而忽视船舶涂装专业英语实际应用的弊端，以企业生产及施工文件中常用的词汇和句式为重点提升学生的语言应用能力，真正做到学以致用。在教材的设计上，添加大量的图片、数据和样例，力求使学生用最直观的方式理解词汇的意思，提升阅读速度。

本教材共包括5个项目，分别是：

Project 1 Ship Corrosion and Electrochemical Protection	Task 1.1 Characteristics and Classification of Metal Corrosion
	Task 1.2 Introduction of Ship Anti-corrosion Methods
Project 2 Ship Coatings	Task 2.1 Characteristics of Ship Coatings
	Task 2.2 Shop Primer
	Task 2.3 Rust-Proof Coatings
	Task 2.4 Finish Coatings
	Task 2.5 Liquid Tank Coatings
Project 3 Pretreatment Before Painting	Task 3.1 Steel Pretreatment Assemble Line
	Task 3.2 Spray Abrasive Surface Treatment
	Task 3.3 Rust Removal by Chemicals
	Task 3.4 Power Tool Cleaning
Project 4 Ship Painting	Task 4.1 Preparation Before Painting
	Task 4.2 Brush Painting and Roller Painting
	Task 4.3 Compressed Air Spraying
	Task 4.4 High-Pressure Airless Spraying

Project 5 Coating Quality Inspection and Control	Task 5.1 Introduction of Common Defects on Coating Surface 1
	Task 5.2 Introduction of Common Defects on Coating Surface 2
	Task 5.3 Field Detection Methods and Steps of Coating Performance

项目一主要介绍船舶腐蚀和电化学保护相关的专业词汇及句式，包括金属腐蚀（metal corrosion）的特点和常见类型等词汇，介绍了电化学保护的两种方式——阳极保护（anodic protection）和阴极保护（cathodic protection）。另外，阴极保护中的牺牲阳极阴极保护法（sacrificial anode cathodic protection method）方面的词汇和句型是学习的重点。项目二的主要内容是船舶涂料（ship coatings）相关的专业词汇和句式，包括车间底漆（shop primer）、防锈涂料（rust-proof coatings）、面层涂料（finish coatings）和液舱涂料（liquid tank coatings）。同时，分别介绍了每种涂料的用途、特点和常见类型。项目三介绍的是与涂装前的预处理（pretreatment before painting）方法相关的专业词汇和句式，包括钢材预处理流水线（steel pretreatment assemble line）、喷射磨料表面处理（spray abrasive surface treatment）、化学方法除锈（rust removal by chemicals）以及动力工具清理（power tool cleaning）。分四个任务介绍了每种处理方法所用工具、处理流程和施工中的注意事项。项目四的主要内容是船舶涂装（ship painting）的主要方法，包括涂装前的准备工作（preparation before painting）、刷涂（brush painting）、辊涂（roller painting）、压缩空气喷涂（compressed air spraying）和高压无气喷涂（high-pressure airless spraying），使学生了解涂装工具、涂装设备以及操作原则等方面的专业词汇和用法。项目五主要介绍了与涂层质量检验与控制（coating quality inspection and control）相关的专业词汇和句式，其中包括涂层缺陷（coating defects）、涂层质量（coating quality）和涂装性能检测（coating performance detection）方法，还介绍了流挂（sagging）等典型涂层缺陷的成因和解决方法。另外还包括干湿膜厚度测量（coating dry/wet film thickness measurement）、粗糙度检验（roughness detection）、附着力测试（coating adhesion test）和钢板表面水溶性盐分的测量（detection of water-soluble salt on the substrate surface）等方面的专业词汇和用法。

本教材由渤海船舶职业学院船舶工程系船舶涂装工程技术专业负责人曹雪主编，负责总体内容的编写、设计和修订；船舶工程系主任王宏老师负责审阅。在编写过程中也得到了渤海船舶重工集团等多位专家和技术人员的鼎力支持及指导，在此谨致谢忱。

由于编者水平有限，难免有疏漏和不当之处，恳请广大读者在使用过程中给予批评指正。

编 者

2022.7.20

目录 Contents

Project 1　Ship Corrosion and Electrochemical Protection ·········· 01
　Task 1.1　Characteristics and Classification of Metal Corrosion ·········· 02
　Task 1.2　Introduction of Ship Anti-corrosion Methods ·········· 08

Project 2　Ship Coatings ·········· 15
　Task 2.1　Characteristics of Ship Coatings ·········· 16
　Task 2.2　Shop Primer ·········· 22
　Task 2.3　Rust-Proof Coatings ·········· 27
　Task 2.4　Finish Coatings ·········· 31
　Task 2.5　Liquid Tank Coatings ·········· 35

Project 3　Pretreatment Before Painting ·········· 40
　Task 3.1　Steel Pretreatment Assemble Line ·········· 41
　Task 3.2　Spray Abrasive Surface Treatment ·········· 48
　Task 3.3　Rust Removal by Chemicals ·········· 53
　Task 3.4　Power Tool Cleaning ·········· 58

Project 4　Ship Painting ·········· 63
　Task 4.1　Preparation Before Painting ·········· 64
　Task 4.2　Brush Painting and Roller Painting ·········· 70
　Task 4.3　Compressed Air Spraying ·········· 74
　Task 4.4　High-Pressure Airless Spraying ·········· 78

Project 5　Coating Quality Inspection and Control ·········· 83
　Task 5.1　Introduction of Common Defects on Coating Surface 1 ·········· 84
　Task 5.2　Introduction of Common Defects on Coating Surface 2 ·········· 91
　Task 5.3　Field Detection Methods and Steps of Coating Performance ·········· 98

参考文献 ·········· 105

Project 1　Ship Corrosion and Electrochemical Protection

Background

The steel ships which sail for long-term in the ocean for a long time will be corroded by various corrosive media inordinately. The corrosion will reduce the strength of the ship structure and bring large damage to the ship. When the corrosion of steel ships reaches a certain extent, the strength of the hull will not fall enough to withstand the huge impact of the hull structure given by the ocean waves, thus causing the inevitable occurrence of shipwreck accidents. When the corrosion of various machines and outfitting equipment on the ship reaches a certain extent, the equipment will not work and operate normally, causing various kinds of mechanical faults. In serious cases, the ship will get out of control and lose the ability to self-help when it is sailing in the sea, even resulting in disaster. Therefore, the ship corrosion to a certain extent can only be scrapped.

Since corrosion is inevitable, through the study of this project, we should understand the corrosion characteristics of the ship in the ocean, master the methods of ship anti-corrosion, and learn to use certain means to control the corrosion speed of the ship, so as to achieve the purpose of extending the service life of the ship.

Learning Target

Knowledge Target

1. Understand the basic characteristics and types of metal corrosion and related vocabulary and sentence patterns.

2. Master the characteristics and importance of ship corrosion and protection and related vocabulary and sentence patterns.

3. Master the main methods and differences of electrochemical anti-corrosion and painting anti-corrosion and related vocabulary and sentence patterns.

Ability Target

1. Understand the design and construction principles of the electrochemical protection system.

2. Master the oral expressions related to the application scope and precautions of the sacrificial anode cathodic protection system.

Task Definition

Task 1.1 Characteristics and Classification of Metal Corrosion
Task 1.2 Introduction of Ship Anti-corrosion Methods

Task 1.1 Characteristics and Classification of Metal Corrosion

Text Reading

Metal materials are the most widely used engineering materials in contemporary society. They are not only used in industrial and agricultural production and scientific research, but also can be seen everywhere in daily life.

Corrosion refers to the deterioration and damage gradually caused under the action of the surrounding ambient media. As the most common form of metal damage, corrosion has been widely valued by countries around the world. The reason is that both large engineering structures and small parts will use metal and contact with the surrounding medium. Some matters will affect the corrosion speed of the metal, such as medium composition, temperature, pressure, pH, and force of the material. It illustrates the corrosion status of the typical ship structural parts. See Figure 1.1, in addition to these external factors, the composition, crystal, structural state and other internal factors of metal materials will also affect the type and the speed of corrosion.

Figure 1.1　The image of corrosive metals

Almost all materials degrade over time once exposed to the natural environment. When iron or steel is exposed to the air and water, we will see that rust generates slowly within hours, sometimes corrosion can occur within minutes, and reddish-brown iron oxide will appear on the surface of iron or steel. In fact, no matter the surface layer is oxide, carbonate, or other compounds, covering the steel surface can help to prevent the surface corrosion. Especially if the surface layer can effectively separate the metal from the environment, it can effectively protect the steel plate. However, ordinary iron or steel plate cannot form such an effective barrier. Whether iron oxide or rust, the structure is loose and porous. It can be easily penetrated by oxygen and moisture, so the steel plates will be rusted continually. Without effective protection, the metal will be eventually corroded and punctured.

According to the production mechanism, metal corrosion can be divided into two categories: chemical corrosion and electrochemical corrosion. According to the destructive form of corrosion, it can also be divided into comprehensive corrosion and local corrosion. Comprehensive corrosion includes uniform comprehensive corrosion and uneven comprehensive corrosion. Local corrosion includes galvanic corrosion, pitting corrosion, crevice corrosion, inter-crystal corrosion, stress corrosion, fatigue corrosion, selective corrosion, empty bubble corrosion, etc.

Electrochemical corrosion is caused by the electrochemical action of the metal surface and the ion conductive medium, including the cathode and anode regions. It is characterized by a current flowing between the metal and the medium. It is the most common and prevalent type of metal corrosion. See Figure 1.2 for specific procedures. Electrochemical corrosion can be divided into:

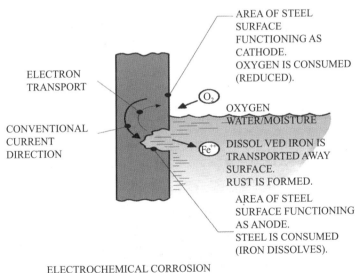

Figure 1.2 The image of electrochemical corrosion

(1) Corrosion in natural water: It mainly includes corrosion in fresh water and seawater. The main sources of fresh water are rainfall, snowmelt or natural spring water. Due to the influence of salt content, conductivity, oxygen content, pH value, temperature, flow rate, marine life, etc. , in sea water, the corrosion rate of metals is much higher than that in fresh water.

(2) Soil corrosion: The corrosion of underground oil, gas and water pipelines, cables, building foundations and other underground works.

(3) Atmospheric corrosion: Corrosion of metals in nature due to the contact with humid air.

(4) Microbial corrosion: Microbial corrosion is not the erosion effect of microorganisms themselves on metals, but the result of their life activities indirectly affects the electrochemical process of metal corrosion (such as anode and cathode reactions, the acid and alkalinity that change the corrosion environment, etc.).

(5) Marine biological corrosion: Many marine organisms adhere to metal equipment in the ocean. They have the production of a large number of corrosive substances during growth, reproduction and metabolism. Meanwhile, due to the uneven coverage of metal surfaces, marine organisms will form oxygen concentration batteries which will also accelerate corrosion to varying degrees.

Comprehensive corrosion is one of the most common forms of corrosion. It occurs on the whole surface of the metal in contact with the medium. There is comprehensive corrosion such as steel corrosion in the atmosphere and seawater or oxidation under high temperature conditions. Another corrosion form corresponding to comprehensive corrosion is local corrosion which is a kind of severe corrosion occuring on the local area of the metal surface, while other parts of the surface are not damaged or have relatively small corrosion damage. See Figure 1.3 for the comparison diagram of comprehensive corrosion and local corrosion.

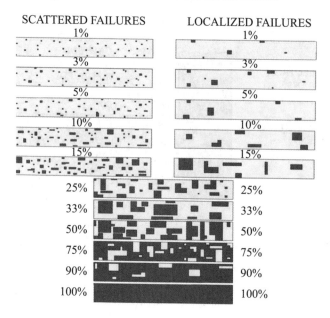

Figure 1.3　Comparison diagram of comprehensive corrosion and local corrosion

Although the area and the amount of corrosion on metal surface is small, local corrosion often becomes the most harmful corrosion type in engineering technology application due to the sudden of corrosion damage and unpredictability of damage time.

Project 1 Ship Corrosion and Electrochemical Protection

New Words and Expressions

corrosion [kəˈrəʊʒn] *n.* 腐蚀，侵蚀，锈蚀
protection [prəˈtekʃn] *n.* 保护
material [məˈtɪərɪəl] *n.* 材料，原料；*adj.* 物质的
scientific research 科学研究
deterioration [dɪˌtɪərɪəˈreɪʃən] *n.* 恶化，变坏；衰退
damage [ˈdæmɪdʒ] *v.* 损坏；伤害
medium [ˈmiːdɪəm] *n.* 方式；材料；介质
degrade [dɪˈɡreɪd] *v.* （使）退化，降解
iron [ˈaɪən] *n.* 铁器，铁制品；*adj.* 铁制的，含铁的
expose [ɪkˈspəʊz] *v.* 使显露；揭露；使暴露于
iron oxide 氧化铁
electrochemistry [ɪˌlektrəʊˈkemɪstrɪ] *n.* 电化学
galvanic corrosion 电偶腐蚀
pitting corrosion 点蚀
crevice corrosion 缝隙腐蚀
inter-crystal corrosion 晶间腐蚀
stress corrosion 应力腐蚀
fatigue corrosion 腐蚀疲劳
selective corrosion 选择性腐蚀
atmospheric [ˌætməsˈferɪk] *adj.* 大气的；大气引起的

Notes

1. Corrosion refers to the deterioration and damage gradually caused under the action of the surrounding ambient media. As the most common form of metal damage, corrosion has been widely valued by countries around the world.
   Translation: 腐蚀是指在周围环境介质的作用下逐渐产生的变质和破坏。作为金属最常见的一种损坏形式，腐蚀已被世界各国广泛重视。

2. Electrochemical corrosion is caused by the electrochemical action of the metal surface and the ion conductive medium, including the cathode and anode regions. It is characterized by a current flowing between the metal and the medium.
   Translation: 电化学腐蚀是由金属表面与离子导电性介质发生电化学作用引起的，在作用过程中有阴极区和阳极区。其特点是在金属与介质中有电流流动。

Expanding Reading

Electrochemical corrosion is the corrosion phenomenon of metal due to the electrochemical reaction in the electrochemical solution medium. The corrosion current is produced during the electrochemical reaction. Marine environment is very harsh corrosive environment, so ships with steel as the main structural material are always facing corrosion hazards in the marine atmosphere and seawater electrolyte.

Steel will form a thin water film on its surface in wet air, which causes electrochemical corrosion on the surface. The product of steel corrosion is the hydrate (rust) of iron's oxide. Its loose texture can not isolate the continuous contact in steel, oxygen and water, therefore, the corrosion will continue to develop in wet air.

The position between near the light and heavy-load water lines and the inter-immersion part of the ship is in similar condition to the splash area of marine fixed steel structure. The corrosion environment of this area is extremely particular. The surface is affected by the periodic wetting and the impact of wind and waves, and it is in an alternating dry and wet state for a long time. As the wave rolls, the speed with oxygen supplied on the steel surface is accelerated. Thus it accelerates the cathode process of corrosion. In addition, the ferrous in the corrosion product suffers from strong oxidation to ferric iron. The whole process can not only inhibit the corrosion, but also lead to cathode depolarization due to the reduction of the ferric iron, which makes the corrosion intensify. Therefore, the splash zone is corroded quite fast.

The underwater part of the ship is complicated by the corrosion environment due to the influence of seawater salinity, dissolved oxygen, temperature, seawater flow rate and marine organisms.

Exercises

I. Answer the following questions according to the passage.

1. Can you describe metal corrosion in your own words?

2. How many kinds of local corrosion do you know?

Ⅱ. Practice these new words.

1. English to Chinese.

corrosion _____   damage _____

electrochemistry _____   expose _____

2. Chinese to English.

腐蚀与防护 _____   点蚀 _____

材料 _____   铁 _____

Ⅲ. Translation.

1. Translate the following sentences into Chinese.

(1) Some matters will affect the corrosion speed of the metal, such as medium composition, temperature, pressure, pH, and force of the material.

(2) Whether iron oxide or rust, the structure is loose and porous. It can be easily penetrated by oxygen and moisture, so the steel plates will be rusted continually. Without effective protection, the metal will be eventually corroded and punctured.

2. Translate the short passage.

According to the production mechanism, metal corrosion can be divided into two categories: chemical corrosion and electrochemical corrosion. According to the destructive form of corrosion, it can also be divided into comprehensive corrosion and local corrosion. Comprehensive corrosion includes uniform comprehensive corrosion and uneven comprehensive corrosion. Local corrosion includes galvanic corrosion, pitting corrosion, crevice corrosion, inter-crystal corrosion, stress corrosion, fatigue corrosion, selective corrosion, empty bubble corrosion, etc.

Task 1.2 Introduction of Ship Anti-corrosion Methods

Text Reading

In fact, the most important mechanical corrosion occurring in the tanks and holds is electrochemical corrosion. The premises are that water exists, current is produced, and dissolved ions are generated. The specific corrosion situation of the ship varies greatly according to the corrosion environment of the hull, the sailing area of the ship, the age and the maintenance degree of the ship. Here we will focus on the corrosion of various parts of the hull in their corrosive environment.

1. Corrosion of the underwater part and waterline area of the hull

Seawater contains a variety of organisms, suspended sediment, dissolved gases, corrupt organic matter, etc., thus the corrosion in the underwater part is more complex than in a pure salt solution. Due to the action of tide and wave rolling, forming the state of dry and wet alternately, the waterline area is the fastest corrosion area (Figure 1.4).

Figure 1.4 Waterline area

The bow of the ship mainly has a large hydrodynamic effect on the hull, especially for ships with relatively high speed. In addition, a huge amount of rolling foam will produce collision on the hull surface, and the impact of anchor chains and floating objects will damage the coating.

The midship is mainly easily damaged when it is relied on the dock.

Due to the strong current, the stern is caused corrosion damage by the high-speed rotation of the propeller (Figure 1.5). The propellers are made of copper alloy, and long-term contact with the iron hull is easily caused anodic polarization.

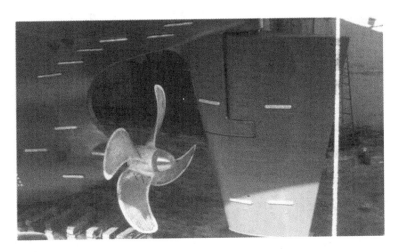

Figure 1.5 Structure of stern

At the bottom of the ship, due to a large number of marine organisms attached, it is easy to produce oxygen concentration batteries (Figure 1.6). In addition, the weld at the bottom of the ship is also a relatively weak corrosion resistance part.

Figure 1.6 Marine organisms attached to the bottom of the ship

2. Corrosion of the hull upper structure

The abundance of chloride in the marine atmosphere aggravates the corrosion of the condensate to the structure. The water splashes into the hull superstructures and dries, leaving a thin layer of salt on the surface to keep the structure wet, thus causing continuous corrosion of the structure (Figure 1.7).

Figure 1.7　Corroded deck structure

　　If the drainage holes on the deck are not arranged or the deck is deformed by external forces, the water will gather in low terrain or where there are no drainage holes. In addition, in some areas on the deck, such as the place above the cabin and the boiler, the high temperature increases the areas which gather water in. (Figure 1.8).

Figure 1.8　Seeper on deck

　　3. Corrosion of the internal structure of the hull

　　The work and residence cabins are protected effectively, therefore, there is usually no obvious structural corrosion.

The top of the sanitary compartment and the lower part of the deck cause severe corrosion damage due to air humidity saturation and long-term water and moisture.

The corrosion condition of the cargo compartment mainly depends on its type of cargo.

In the liquid tank, whether drinking fresh water, oil or other corrosive media, it will be caused corrosion to the cabin. If no better protective measures are taken, it will cause long-term losses and harm to the ship.

The anti-corrosion methods of ships are mainly divided into two kinds. One is electrochemical protection, and the other is coating protection. Electrochemical protection can be divided into anodic protection and cathodic protection. Cathodic protection can be refined into the sacrificial anodic cathodic protection method and external current cathodic protection method. Currently, the most common thing for large shipbuilding enterprises and companies at home and abroad is to sacrifice the anode cathodic protection method. This book will introduce coating protection in later sections.

New Words and Expressions

tank [tæŋk] *n.* 船舱
hold [həʊld] *n.* 货舱
hull [hʌl] *n.* 船体
maintenance [ˈmeɪntənəns] *n.* 维修保养
waterline [ˈwɔːtəlaɪn] *n.* 水线
organism [ˈɔːɡənɪzəm] *n.* 有机体；微生物
wave [weɪv] *n.* 海浪；波浪
bow [baʊ] *n.* 船头
collision [kəˈlɪʒn] *n.* 碰撞；冲突
anchor [ˈæŋkə(r)] *n.* 锚
stern [stɜːn] *n.* 船尾
propeller [prəˈpelə(r)] *n.* 螺旋桨；推进器
copper alloy 铜合金
bottom [ˈbɒtəm] *n.* 船底
oxygen [ˈɒksɪdʒən] *n.* 氧气
weld [weld] *vt.& vi.* 焊接
drainage hole 排水孔
deck [dek] *n.* 甲板
cabin [ˈkæbɪn] *n.* 机舱
cargo [ˈkɑːɡəʊ] *n.* 货物

Notes

1. Seawater contains a variety of organisms, suspended sediment, dissolved gases, corrupt organic matter, etc., thus the corrosion in the underwater part is more complex than in a pure salt solution.
   Translation: 海水中含有多种生物、悬浮泥沙、溶解的气体、腐败的有机物等，因此水下部分的腐蚀比在纯的盐溶液里要复杂得多。
2. The anti-corrosion methods of ships are mainly divided into two kinds. One is electrochemical protection, and the other is coating protection. Electrochemical protection can be divided into anodic protection and cathodic protection.
   Translation: 船舶的防腐方法主要分为两种，一是电化学保护，二是涂层防护。电化学保护又可分为阳极保护和阴极保护。

Expanding Reading

In general, the ship usually paints the coatings after the surface treatment. The coating can isolate the harm to steel well. However, the complete ideal coating does not exist. Due to the construction process of transportation, welding, installation, coating aging or small pinholes in the coating, there are always certain defects on the external coating of the metal surface. This problem can be solved effectively if we combine electrochemical protection technology with coatings. The cost of cathodic protection only accounts for 1%~3% of the cost of the protected metal structures, the service life of the structures can be doubled or even dozens of times longer.

Electrochemical protection is a method of reducing the corrosion speed of metal by changing the metal potential to the corrosion-free zone or passivation region in the metal potential-pH diagram.

The cathodic protection method in the electrochemical protection method was adopted before the establishment of the electrochemical science. In 1824, British scientist, Humphrey Davy, applied the concept of the sacrificial anode cathodic protection method to naval ships at first time, protecting the copper layer wrapped outside the wooden hull with cast iron, and effectively preventing copper corrosion.

The principle of cathodic protection is to minimize the anodic corrosion rate of dissolution by applying a cathodic current to the protected structure. If the cathodic current is transmitted to the metal by the applied power supply, it is called the applied current cathodic protection method; if the cathodic current is provided by another metal with a more negative potential, it is called the sacrificial anode cathodic protection method (Figure 1.9).

Figure 1.9　The corroded sacrificial anode in ballast tank

Exercises

I. Answer the following questions according to the passage.

1. Can you describe characteristics of ship corrosion in your own words?

2. Can you briefly introduce the corrosion characteristics and protection methods of the lower splash area?

II. Practice these new words.

1. English to Chinese.

   waterline area _____　　hull _____

   moisture _____　　sacrificial anode _____

2. Chinese to English.

   电化学保护 _____　　飞溅区 _____

   压载水舱 _____　　防污性 _____

III. Translation.

1. Translate the following sentences into Chinese.

(1) The specific corrosion situation of the ship varies greatly according to the corrosion environment of the hull, the sailing area of the ship, the age and the maintenance degree of the ship. Here we will focus on the corrosion of various parts of the hull in their corrosive environment.

(2) The abundance of chloride in the marine atmosphere aggravates the corrosion of the condensate to the structure. The water splashes into the hull superstructures and dries, leaving a thin layer of salt on the surface to keep the structure wet, thus causing continuous corrosion of the structure.

2. Translate the short passage.

The anti-corrosion methods of ships are mainly divided into two kinds. One is electrochemical protection, and the other is coating protection. Electrochemical protection can be divided into anodic protection and cathodic protection. Cathodic protection can be refined into the sacrificial anode cathodic protection method and external current cathodic protection method. Currently, the most common thing for large shipbuilding enterprises and companies at home and abroad is to sacrifice the anode cathodic protection method. This book will introduce coating protection in later sections.

Project 2 Ship Coatings

Background

Paint is usually a viscous liquid with mobility, applied to the surface, dried and hardened under room temperature or heating, forming a tough and elastic skin film on the surface of the object to protect, decorate or give special effects to the material. It is divided into the following four categories by its major functions.

1. Protective function

The coatings isolate metal from the atmosphere and other media, preventing oxygen, water vapor, carbon dioxide, sulfur dioxide and other substances in the atmosphere and other corrosive environment from contacting with metals, playing a role in preventing corrosion and extending the service life of the metal.

2. Decoration function

It is pleasing for coatings to see with its smooth or artistic appearance and brilliant and colorful decorative objects and environment.

3. Mark function

A variety of different colors of paint painted on a variety of different occasions can give people eye-catching signs. For example, red fire equipment; orange red ship life-saving equipment (Figure 2.1); various pipes in the cabin painted with specific colors, etc.

Figure 2.1 The lifeboat on the ship

4. Special function

Fire-proof coatings can prevent and stop the spread of flame; heat resistant coatings can protect metal at high temperature; sound absorbing coatings can reduce noise; damping coatings can decrease vibration; reflective coatings can make striking tips at night; camouflage coatings can confuse the real with the false, etc.

Learning Target

Knowledge Target

1. Understand the basic characteristics and types of ship coatings.
2. Master the role of coatings with different functions.

Ability Target

1. Understand the characteristics and classification of ship coatings.
2. Master the varieties of shop primer and know how to choose them.
3. Master the varieties of rust-proof coatings and know how to choose them.
4. Master the varieties of finish coatings and know how to choose them.
5. Master the varieties of liquid cabin coatings and konw how to choose them.

Task Definition

Task 2.1   Characteristics of Ship Coatings
Task 2.2   Shop Primer
Task 2.3   Rust-Proof Coatings
Task 2.4   Finish Coatings
Task 2.5   Liquid Tank Coatings

Task 2.1   Characteristics of Ship Coatings

Text Reading

Ship coatings refer to a general term of various coatings painted on the internal and external surfaces, structural parts and equipment that can extend the service life of the ship and meet the corrosion prevention requirements of the ship. Because the ships are always working in special environment and the ship coatings are different from other steel structures, there are certain special requirements for the ship coatings. Generally speaking, there are several following characteristics for ship coatings:

(1) Thick film-type coatings are preferred. Some areas of the ship are difficult to build, so it is better to paint a high film thickness.

(2) It is suitable for high-pressure airless spraying. Due to the large construction area of ship coatings, the coatings should be suitable for high pressure without gas spraying with high construction efficiency.

(3) It can be dried down at room temperature. Due to the large painting area of the ship, the ship coatings are required to dry at room temperature, requiring heat to dry, or other auxiliary means to dry the coatings which are not suitable for application.

(4) Coatings applied to underwater parts shall be alkaline resistance and potential resistance. Because the underwater part of the ship is often cathode protected, and the surrounding water is alkaline. The coatings should be alkaline resistance and potential resistance.

(5) The coatings applied in the interior of the cabin and the superstructure should not be easy to be burned, and will not emit excessive smoke once burn. This is from the perspective of fire safety which aims to avoid safety accidents.

Each part of the ship is in different environment, so the requirements for coatings are different. The characteristics of the coatings selected according to the ship location are listed below.

1. Atmospheric exposure area

The freeboard, superstructure exterior, open deck and deck outfitting parts are in the marine atmospheric exposure area. These parts are in the humid marine atmosphere containing salt for a long time, suffering from sun exposure and wave impact, so the coatings used are required to have excellent abilities of rust resistance, weather resistance and impact resistance. In addition, it is also to be a good color and light protection.

2. Waterline area

Due to long-term waterline water immersion, erosion and sunlight exposure of dry and wet alternating state, it leads to serious corrosion. Before the coatings used for waterline area, it must have good functions of water resistance, weather resistance, dry and wet resistance, and coating friction, which should be impact-resistant and tough. If the ship adopts cathodic protection, it will have a good function of alkaline resistance.

3. Bottom area

The bottom area has been long-term submerged in seawater, suffering from the electrochemical corrosion of the seawater and damaged by marine organisms. If cathodic protection is adopted, the coatings used in the ship bottom area must have good functions of water resistance, alkaline resistance and wear resistance. The outermost coatings should also be required to have an ability of the anti-fouling to prevent the attachment to marine organisms.

4. Liquid compartment

The liquid tank inside the ship mainly includes ballast tank, drinking tank, fuel tank, sliding oil tank and cargo oil tank of the oil tanker.

The ballast water tank is in the alternating state of dry or wet between seawater load and empty load for a long time. The environment is hot and humid, high salt, no ventilation. Thus the conditions are harsh and it is difficult to maintain. The coatings are required to have excellent abilities of water resistance, salt mist resistance, alternating dry and humidity resistance and excellent corrosion resistance.

Drinking water tank (including fresh water tank) stores drinking water and fresh water for a long time. On the one hand, there is a good water resistance function of coatings. On the other hand, because the drinking water is directly related to the health of the human body, the drinking water tank coatings must absolutely ensure that it will not pollute the water quality. So it must be recognized by the relevant health departments.

Fuel tank or sliding oil tank is used for long-term storage of fuel oil and slid oil, generally it is not easy to be corroded. In principle, it is not allowed to be painted. However, the surface must be cleaned before sealing the capsule, and protected with the corresponding oil film. If the cargo tank of the oil tanker is loaded alternately of oil and water, the coatings selected should not only have a good oil resistance ability, but also have a good function of water resistance and alternate loading performance.

5. Engine room and pump compartment

Engine room and pump compartment are the main workplaces for ships. The indoor temperature is higher than the general cabin interior. The roof and bulkhead coatings of the machine and pump compartment are not easy to be burned, and no excessive smoke will be released once they are burning. So the selected coating varieties need to be approved by the ship inspection department. Oil and water often accumulate at the bottom of the machine and pump compartment, thus the selected coatings are required to have good oil resistance and water resistance abilities.

6. Cabin

The decoration in the cabin is generally not painted. However, the rust-proof coatings should still be painted under the insulation layer and inside the decorative plate. Some work cabins, storage rooms, and health offices still need to be protected and decorated with coatings. Therefore, the coatings used for the cabin should have a good function of rust resistance, and the surface coatings should be decorated by good decoration. In order to ensure fire safety, the coatings should not be easy to burn. It will not release excessive smoke once the combustion happens, so the selected coatings also need to be recognized by the ship inspection department.

New Words and Expressions

coating [ˈkəʊtɪŋ] *n.* 涂层；涂料

structural [ˈstrʌktʃərəl] *adj.* 结构（上）的
thick film-type coating 厚膜型涂料
high-pressure airless spraying 高压无气喷涂
alkaline [ˈælkəlaɪn] *adj.* 碱性
cathodic protect 阴极保护
interior [ɪnˈtɪərɪə(r)] *n.* 内部
superstructure [ˈsuːpəstrʌktʃə(r)] *n.* 上层建筑
burn [bɜːn] *v.* 燃烧
freeboard [ˈfriːˌbɔːd] *n.* 干舷
outfitting [ˈaʊtˌfɪt] *n.* 舾装；舾装件
immersion [ɪˈmɜːʃn] *n.* 浸没
friction [ˈfrɪkʃn] *n.* 摩擦
anti-fouling 防污性
ballast tank 压载舱
ventilate [ˈventɪleɪt] *vt.* 通风
salt mist 盐雾
sealing [ˈsiːlɪŋ] *v.* 密封
engine room 机舱
pump [pʌmp] *n.* 泵

Notes

1. Ship coatings refer to a general term of various coatings painted on the internal and external surfaces, structural parts and equipment that can extend the service life of the ship and meet the corrosion prevention requirements of the ship.
    Translation: 船舶涂料是指涂装于船舶内外表面、结构件以及设备上的，能够延长船舶使用寿命和满足船舶防腐蚀要求的各种涂料的统称。
2. The ballast water tank is in the alternating state of dry or wet between seawater load and empty load for a long time. The environment is hot and humid, high salt, no ventilation. Thus the conditions are harsh and it is difficult to maintain.
    Translation: 压载水舱长期处于海水压载和空载的干湿交替状态，环境湿热，盐分高，密不通风，条件恶劣，且维修难度大。

Expanding Reading

The coatings are generally composed of non-volatile and volatile components.

Non-volatile component is the composition of coatings, also known as the solid part of coatings. It includes the main film forming material, secondary film forming material and auxiliary film forming material. The volatile components only exist in the coatings, and gradually evaporate during the film-forming process of the painting. Finally, it will not exist in the coating.

The main film material can be a separate film. It can also bond paint, filler and other secondary film material film together. It is the main body of coatings and also the basis of coatings. Therefore, it is called the "base material".

Secondary film-forming substances, namely pigment and filler, are present in the color paint. It cannot leave the main film material alone, but it is a more important component of the paint film. The role of secondary film material is to improve the performance of paint film.

Auxiliary film forming material can not be formed alone. Its role is to improve the construction performance and film forming performance of the coatings, so it is called auxiliary agent.

Volatile component refers to the solvent, which acts to make the coatings into liquid for construction, and to adjust the viscosity of the coatings and meet the requirements suitable for construction. Once construction is completed, its mission is over and it should be left the paint, otherwise the paint will not be formed or dried completely.

Exercises

Ⅰ. Answer the following questions according to the passage.

1. What are the characteristics of the ship coatings?

2. Why the oil tank can not be painted?

Ⅱ. Practice these new words.

1. English to Chinese.
   coating _____   airless spraying _____
   anti-fouling _____   ballast tank _____

2. Chinese to English.
   密封 _____   干舷 _____
   舾装 _____   上层建筑 _____

Ⅲ. Translation.

1.Translate the following sentences into Chinese.

(1) Due to the large construction area of ship coatings, the coatings should be suitable for high pressure without gas spraying with high construction efficiency.

(2) If the cargo tank of the oil tanker is loaded alternately of oil and water, the coatings selected should not only have a good oil resistance ability, but also have a good function of water resistance and alternate loading performance.

2. Translate the short passage.

The decoration in the cabin is generally not painted. However, the rust-proof coatings should still be painted under the insulation layer and inside the decorative plate. Some work cabins, storage rooms, and health offices still need to be protected and decorated with coatings. Therefore, the coatings used for the cabin should have a good function of rust resistance, and the surface coatings should be decorated by good decoration. In order to fire safety, the coatings should not be easy to burn. It will not release excessive smoke once the combustion happens, so the selected coatings also need to be recognized by the ship inspection department.

Task 2.2 Shop Primer

Text Reading

Shop primer, also known as a maintenance primer or pretreatment primer, is a primer used on the assembly line for steel plate or type steel after blasting pretreatment. The function of shop primer is to protect the surface of blasting steel. It can prevent rust during the processing and assembly. Its function is to minimize the workload of surface treatment during segmentation or platform painting.

As a temporary protective primer, the main role of shop primer is to play a temporary protection of the steel surface from corrosion before painting. It can be removed or retained before the formal painting, mainly depending on the integrity of the shop primer and the specific surface treatment requirements of the first coat in the formal painting.

Since the shop primer can not be removed before the formal painting, the existence of the shop primer should not only have no adverse impact on welding and cutting, but also be used with various ship coatings. In addition, the shop primer should be painted immediately after the surface treatment, and it must form thin film on an automatic assembly line. So it is required to dry within 5 minutes at normal temperature (23℃).

The general dry film thickness of shop primer is from 15 to 25 μm, preventing rust for at least 3 months in the marine or industrial atmosphere. Moreover, the film thickness of the shop primer is generally not included in the total film thickness of the hull coating.

In addition, because the shop primer in welding, cutting, fire engineering correction that film heat damage area is small, it should be good impact resistance and toughness in order to be suitable for the machining of painted steel.

There are four categories of commonly used shop primers, which are phosphate primer, epoxy zinc-rich primer, epoxy zinc-free primer and inorganic zinc primer.

1. Phosphate primer

The main components are zinc chromate, polyvinyl alcohol, butyl aldehyde and phenolic resin. The standard film thickness is from 5 to 15 μm, the outdoor corrosion prevention period is from 3 to 4 months, and the general dry performance is about 5 minutes at 23 ℃. It has several functions of excellent weldability performance, excellent cutting performance, small environmental pollution, poor heat resistance, poor potential resistance, excellent construction performance and general solvent resistance. There is an excellent performance with the rust-proof paint for oil, ethylene and polyurethane.

2. Epoxy zinc-rich primer

The main components are zinc powder, epoxy resin and polyamide resin. The standard film thickness is from 15 to 20 μm, and the outdoor corrosion prevention period is from 6 to 9 months. Its drying performance is about 5 minutes at 23 ℃. Its weldability performance and the cutting performance are in general level, however its environmental pollution is serious. The ZnO concentration is so large that it is easy to exceed the standard. It has several functions such as good heat resistance, excellent potential resistance, general construction performance, and good solvent resistance. There is an excellent performance with asphalt, chloride rubber, epoxy asphalt and pure epoxy rust-proof paint, but it can not be used with oily paint.

3. Epoxy zinc-free primer

The main components are red iron oxide, epoxy resin, and polyamide resin. The standard film thickness is from 20 to 25 μm, and the outdoor corrosion prevention period is 4 months. Its drying performance is about 5 minutes at 23 ℃. Its weldability and cutting performance are at good levels with small environmental pollution, however, its heat resistance is poor and potential resistance is at general level. It is good at construction performance and good solvent resistance. Performance matching with various types of rust-proof paint is good.

4. Inorganic zinc primer

The main components are zinc powder and ethyl orthosilicate. The standard film thickness is from 15 to 25 μm, the outdoor corrosion prevention period is from 6 to 9 months, and the drying speed is fast, about 1 to 2 minutes at 23 ℃. With its excellent welding ability and good cutting performance, there is a little environmental pollution and it can produce a certain amount of ZnO. But the amount usually does not exceed the standard. Its thermal resistance and potential resistance are excellent, and construction performance is at general level. But its solvent resistance is also excellent. The antirust paint with epoxy asphalt is at excellent level but it can not be used with oily paint.

New Words and Expressions

primer ['praɪmə(r)] *n.* 底漆
blasting ['blɑːstɪŋ] *n.* 抛丸
pretreatment [priːˈtriːtmənt] *n.* 预处理
prevent [prɪˈvent] *v.* 阻碍；预防
segmentation [ˌsegmenˈteɪʃn] *n.* 分段；分节
platform ['plætfɔːm] *n.* 船台；平台
integrity [ɪnˈtegrətɪ] *n.* 完整性
automatic [ˌɔːtəˈmætɪk] *adj.* 自动的；自动化的
film thickness 漆膜厚度

phosphate primer 磷化底漆
rust-proof paint 防锈漆
epoxy zinc-rich primer 环氧富锌底漆
resin ['rezɪn] *n.* 树脂；合成树脂
chloride rubber 氯化橡胶
epoxy zinc-free primer 环氧无锌底漆
inorganic zinc primer 无机锌底漆

Notes

1. Shop primer, also known as a maintenance primer or pretreatment primer, is a primer used on the assembly line for steel plate or type steel after blasting pretreatment.
   Translation: 车间底漆又称保养底漆或预处理底漆，是钢板或型钢经抛丸预处理后在流水线上使用的一种底漆。
2. Since the shop primer can not be removed before the formal painting, the existence of the shop primer should not only have no adverse impact on welding and cutting, but also used with various ship coatings.
   Translation: 由于在正式涂装前，车间底漆可以不必除去，因此车间底漆的存在应对焊接与切割无不良影响且要能够与各种船舶涂料配套使用。

Expanding Reading

In addition to the above four shop primers, from the late 1980s to the early 1990s, a new generation of high temperature resistant inorganic zinc shop primer was launched abroad. This new inorganic zinc primer is based on the original inorganic zinc shop primer, using super heat resistance resin to modify ethyl silicate, and a part of heat resistance rust-proof pigment and zinc powder share to reduce the content of zinc powder and improve its heat resistance in the shop primer.

Compared with the traditional shop primer heat resistance, high temperature resistant inorganic zinc shop primer is greatly improved from 400℃ resistance to 800℃ resistance, so that the area of welding and fire correction site coating loss will be greatly reduced. In addition, the reduction of zinc content not only reduces the amount of zinc oxide soot produced in the thermal processing area, but also benefits to the health of workers. It also reduces the amount of white zinc salt on the surface of the primer in the workshop after a period of outdoor exposure. Then the workload of secondary rust removal is reduced greatly. This is of great significance to some shipbuilding countries where the labor is lacking and expensive. Currently, although the price of this high temperature resistant inorganic zinc shop primer is about 30% higher than that of the traditional inorganic zinc shop primer, it has been expanded gradually in some countries.

Exercises

I. **Answer the following questions according to the passage.**

1. Can you describe characteristics of shop primer in your own words?

2. Can you briefly introduce the commonly used varieties of shop primer?

II. **Practice these new words.**

1. English to Chinese.
   primer _____   blasting _____
   pretreatment _____   film thickness _____

2. Chinese to English.
   磷化底漆 _____   环氧富锌底漆 _____
   环氧无锌底漆 _____   无机锌底漆 _____

III. **Translation.**

1. Translate the following sentences into Chinese.

   (1) The function of shop primer is to protect the surface of blasting steel. It can prevent rust during the processing and assembly. Its function is to minimize the workload of surface treatment during segmentation or platform painting.

   (2) In addition, the shop primer should be painted immediately after the surface treatment, and it must form thin film on an automatic assembly line. So it is required to dry within 5 minutes at normal temperature (23℃).

2. Translate the short passage.

The main components are red iron oxide, epoxy resin, and polyamide resin. The standard film thickness is from 20 to 25 μm, and the outdoor corrosion prevention period is 4 months. Its drying performance is about 5 minutes at 23 ℃. Its weldability and cutting performance are at good levels with small environmental pollution, however, its heat resistance is poor and potential resistance is at general level. It is good at construction performance and good solvent resistance. Performance matching with various types of rust-proof paint is good.

Task 2.3 Rust-Proof Coatings

Text Reading

Rust-proof coatings are based on the material itself to isolate the corrosion medium from the steel surface. In addition, the rust prevention effect mainly depends on rust-proof pigments. Rust-proof pigments can be divided into physical rust-proof pigments and chemical rust-proof pigments.

1. Physical rust-proof pigments

Physical rust-proof pigment is a kind of chemical property which is stable relatively with the help of its fine particles filling paint film structure. It can improve the density of paint film and reduce the permeability of paint film in order to achieve a certain rust prevention effect. Commonly used physical rust prevention pigments are red clay, iron red, iron yellow, aluminum powder, mica iron oxide, etc. Among them, the aluminum powder and mica iron oxide are flake structures, which are superimposed in the paint film, and have functions of good waterproof, air permeability and weather resistance.

2. Chemical rust-proof pigments

Chemical rust-proof pigment is a kind of pigment that can rely on its own chemical properties to prevent or reduce the corrosion effect on steel and play a rust prevention effect. Red lead, calcium lead acid, lead cyanide, etc. are used commonly.

The rust-proof coatings in ship paint consist of two parts: bottom rust-proof coatings and marine rust-proof coatings.

(1) Rust-proof coatings on the ship bottom

Rust-proof coatings on the ship bottom shall have the following properties:

① The paint film must have a good function of wet adhesion on the surface of the steel plate, and the paint film will not bubble and fall off under the long-term soak and water flow impact.

② Good drying performance.

③ With the good ability of matching with the anti-fouling coatings of the ship bottom and good layer and layer adhesion, it does not interfere with the anti-fouling performance of the anti-fouling coatings.

④ With the good alkaline resistance and potential resistance, it can be used together with the cathodic protection device.

The common bottom rust-proof coatings are asphalt system, chlorinated rubber system, epoxy asphalt type, ethylene bottom rust-proof coatings, etc.

(2) Marine rust-proof coatings

Marine rust-proof coatings usually refer to the conventional rust-proof coatings. They are used in the atmospheric exposure area and engine room, cabin, cargo compartment and other indoor areas above the waterline, excluding the rust-proof coatings on the bottom and the special rust-proof coatings used in the liquid tank. Marine rust-proof coatings shall have the following properties:

① Good adhesion with the steel surface, and can be matched with all kinds of finish coatings.

② Good water resistance and weather resistance.

③ Good mechanical strength of paint film, and has certain impact resistance and friction resistance, as well as a good toughness.

The common marine rust-proof paint is red lead, chromate rust-proof paint, etc.

New Words and Expressions

isolate [ˈaɪsəleɪt] *vi.* 隔离
physical [ˈfɪzɪkl] *adj.* 物理的
pigment [ˈpɪgmənt] *n.* 颜料；色料
particle [ˈpɑːtɪkl] *n.* 微粒；颗粒
permeability [ˌpɜːmɪəˈbɪlətɪ] *n.* 渗透性
red clay 红土
iron red 铁红
aluminum powder 铝粉
mica iron oxide 云母氧化铁
property [ˈprɒpətɪz] *n.* 特性
red lead 红丹
calcium plumbite 铅酸钙
lead cyanamide 氰氨化铅
adhesion [ədˈhiːʒn] *n.* 黏合；黏附力
anti-fouling coating 防污涂料
asphalt [ˈæsfælt] *n.* 沥青
ethylene [ˈeθɪliːn] *n.* 乙烯

Notes

1. Rust-proof coatings are based on the material itself to isolate the corrosion medium from the steel surface. In addition, the rust prevention effect mainly depends on rust-proof pigments.
Translation: 防锈涂料是依靠基料本身将腐蚀介质与钢铁表面进行隔离，另外，防锈效果主要取决于防锈颜料。

2. Chemical rust-proof pigment is a kind of pigment that can rely on its own chemical properties to prevent or reduce the corrosion effect on steel and play a rust prevention effect.

Translation: 化学防锈颜料是在涂层中能够依靠自身的化学性质阻止或降低腐蚀介质对钢铁的腐蚀作用，起到防锈效果的一类颜料。

Expanding Reading

Complete bottom rust-proof paint system according to GB / T 13351—1992 can be divided into three class by the validity period of rust-proof.

First class: Rust prevention is valid for 30 months.

Second class: Rust prevention is valid for 18 months.

Third class: Rust prevention is valid for 12 months.

The complete system of bottom rust-proof paint must be capable of application and dried under the natural environmental condition. The drying time of each coating shall meet the product standard. The film thickness of each layer must meet the construction specification or technical requirements. And it should be suitable with cathodic protection of the ship.

Exercises

I. Answer the following questions according to the passage.

1. How does rust-proof coatings achieve the antirust effect?

2. What kinds of rust-proof coatings are there?

II. Practice these new words.

1. English to Chinese.

   isolate _____   rust-proof coatings_____

   pigment _____   adhesion_____

2. Chinese to English.

   渗透性_____   防污涂料_____

   黏附力_____   沥青_____

III. Translation.

1. Translate the following sentences into Chinese.

(1) Physical rust-proof pigment is a kind of chemical property which is stable relatively with the help of its fine particles filling paint film structure. It can improve the density of paint film and reduce the permeability of paint film in order to achieve a certain rust prevention effect.

(2) The paint film must have a good function of wet adhesion on the surface of the steel plate, and the paint film will not bubble and fall off under the long-term soak and water flow impact.

2. Translate the short passage.

Marine rust-proof coatings usually refer to the conventional rust-proof coatings. They are used in the atmospheric exposure area and engine room, cabin, cargo compartment and other indoor areas above the waterline, excluding the rust-proof coatings on the bottom and the special rust-proof coatings used in the liquid tank.

Task 2.4 Finish Coatings

Text Reading

The finish coatings are divided into ship shell paint, waterline area finish paint, deck paint, cargo paint, cabin interior finish paint and anti-fouling paint on the ship bottom.

(1) Ship shell paint refers to the finish coatings used for the area above the heavy load waterline out of superstructure and the exposed part of the deck outfitting part which requires several functions such as good weather resistance, certain water resistance and impact resistance, and good decoration.

The commonly used ship shell paint includes alkyd shell paint, chlorinated rubber shell paint, acrylic resin shell paint, epoxy shell paint, etc.

(2) The waterline area finish coatings should have several abilities such as good water resistance, impact resistance, dry and wet resistance of alternating performance, and need a good decorative effect.

The commonly used waterline finish coatings include phenolic waterline coating, chlorinated rubber waterline coating and epoxy waterline coating.

Currently, many ship owners want to use the same coating system as the bottom area and waterline area, so the finish coatings in the waterline area are mostly self-polishing anti-fouling paint.

(3) Deck paint is the finish coating which is used for the deck part of ships. Therefore, it should have several functions such as good adhesion, water resistance, weather resistance and wear resistance.

Common deck paint includes phenolic deck paint, chlorinated rubber deck paint, and epoxy deck paint.

All kinds of deck paint can be added to some anti-skid materials, such as diamond sand, rubber fine particles, and plastic fine particles. These materials are made from deck anti-skid paint and make the paint film surface rough in order to prevent personnel from the wind and waves or wet deck walking fall.

(4) Cargo hold paint is used in the interior of the ship cargo holds. It should be required good adhesion, high hardness and wear resistance. Common cargo hold paint includes phenolic, chlorinated rubber, epoxy, bleach epoxy asphalt, etc.

For the cargo holds loaded with grain, the cargo hold paint should be non-toxic and pollution-free to the grain, and obtain the approval certificate of the relevant departments.

(5) Cabin interior with finish paint mainly refers to the interior surface of the cabin coatings in the engine room. The adhesion and decoration should be good, and not easy to burn. It will

not produce excessive smoke when it is burning. These are the standards for both international maritime life safety convention (SOLAS) and the provisions of many classification societies, so the use of the paint needs to be recognized by the ship inspection department.

Commonly used cabin interior with finish paint includes oil-based resin paint and alkyd resin paint.

(6) Anti-fouling paint for ship bottom is usually called anti-fouling paint which is a special coating to prevent marine life from damaging and to keep the hull clean. It is different from the general coatings and it is composed of poison material, seepage agent, base material, pigment, aid and solvent.

Commonly used anti-fouling paint types are dissolution type, contact type, diffusion type, and hydrolysis type (self-polishing type).

New Words and Expressions

finish coating 面层涂料
heavy load waterline 重载水线
decoration [ˌdekəˈreɪʃn] *n.* 装饰性
alkyd shell paint 醇酸船壳漆
acrylic resin shell paint 丙烯酸树脂船壳漆
phenolic waterline coating 酚醛型水线漆
self-polish 自抛光
anti-skid 防滑
hardness [ˈhɑːdnəs] *n.* 硬度；坚硬
bleach epoxy asphalt 漂白环氧沥青
grain [greɪn] *n.* 谷物
approval [əˈpruːvl] *n.* 认可；认证
excessive [ɪkˈsesɪv] *adj.* 过分的；过度的
poison [ˈpɔɪzn] *n.* 毒料；毒药
seepage agent 渗出助剂

Notes

1. The finish coatings are divided into ship shell paint, waterline area finish paint, deck paint, cargo paint, cabin interior finish paint and anti-fouling paint on the ship bottom.
   Translation: 面层涂料分为船壳漆、水线区域面层涂料、甲板漆、货舱漆、舱室内部用面漆和船底防污漆。

2. Anti-fouling paint for ship bottom is usually called anti-fouling paint which is a special coating to prevent marine life from damaging and to keep the hull clean.

Translation: 船底防污漆通常被称为防污涂料，是防止海洋生物破坏、保持船体光洁的一种专用涂料。

Expanding Reading

There are more than 8,000 species of plants and 59,000 species of marine organisms in the sea, including more than 600 species of attached plants and 18,000 species of attached animals. They often adhere, colonize and breed further on objects, such as shore rocks, underwater structures or underwater boat bottoms.

A large number of marine organisms attach to the bottom of the ship. They will bring great harm to the ship, because they will not only increase the weight of the ship and reduce the load of the ship, but also greatly increase the resistance of the hull and fuel comsumption, and reduce the speed of the ship.

Therefore, the ship bottom anti-fouling paint should have the following characteristics:

(1) It can stop marine organisms adhering to the bottom of the ship within a certain period of time.

(2) The paint film contains a certain amount of toxic materials that can kill the adhering marine organisms, meanwhile it can continuously and gradually seep out into the sea water.

(3) Contrary to the rust-proof coatings, the paint film should have certain permeability to maintain the continuous seepage of toxic material.

(4) There is good adhesion of the rust-proof coatings, and the anti-fouling coatings should have good adhesion among the layers. Each layer should be slightly soluble.

(5) The paint film has good seawater impact resistance, and does not bubble and fall off under long-term water immersion conditions.

(6) After a period of sailing, it should have varying degrees of polishing.

Exercises

I. Answer the following questions according to the passage.

1. Can you briefly introduce the types of finish coatings?

2. How is the anti-fouling paint self-polished?

Ⅱ. Practice these new words.

1. English to Chinese.

　　finish coating _____　　alkyd shell paint _____

　　self-polish _____　　posion _____

2. Chinese to English.

　　重载水线 _____　　防滑涂料 _____

　　漂白环氧沥青 _____　　渗出助剂 _____

Ⅲ. Translation.

1. Translate the following sentences into Chinese.

　　(1) Ship shell paint refers to the finish coatings used for the area above the heavy load waterline out of superstructure and the exposed part of the deck outfitting part which requires several functions such as good weather resistance, certain water resistance and impact resistance, and good decoration.

　　(2) Cargo hold paint is used in the interior of the ship cargo holds. It should be required good adhesion, high hardness and wear resistance.

2. Translate the short passage.

　　Deck paint is the finish coating which is used for the deck part of ships. Therefore, it should have several functions such as good adhesion, water resistance, weather resistance and wear resistance.

　　Common deck paint includes phenolic deck paint, chlorinated rubber deck paint, and epoxy deck paint.

　　All kinds of deck paint can be added to some anti-skid materials, such as diamond sand, rubber fine particles, and plastic fine particles. These materials are made from deck anti-skid paint and make the paint film surface rough in order to prevent personnel from the wind and waves or wet deck walking fall.

Task 2.5 Liquid Tank Coatings

Text Reading

The liquid compartment of ships is divided into two categories: water tank and oil tank. A water cabin includes fresh water tank, drinking water tank, ballast water tank, cooling water tank, bottom water tank, etc. The oil tank includes fuel oil tank, sliding oil tank, dirty oil and water tank, cargo oil tank of crude oil carrier, cargo oil tank of refined oil ship and chemical ship, etc.

1. Water tank coatings

(1) The ballast tank is used for various ballast water tank, bow tip tank, stern tip tank, cabin bottom water tank, etc. Due to the harsh corrosion environment of these tanks, the coatings are required to have excellent water resistance, salt mist resistance, dry and wet alternative resistance and corrosion resistance abilities. In addition, the construction conditions of these parts are harsh, so a higher film thickness is required for the coating to reduce the number of painting times.

In recent years, at the request of ship owners in the world, bleached epoxy asphalt coating without coal tar asphalt is used instead of traditional epoxy asphalt coating in the ballast tank of new shipbuilding. To a certain extent, the coating production personnel and coating construction personnel are away from the harm of coal tar asphalt. Meanwhile, because these coatings can make the cabin color bright and light, it is easy to find the disadvantages of the coatings. It is conducive to the control of the construction quality.

As for the technical requirements of ballast water tank coating, the new PSPC standard is mainly implemented.

(2) Drinking tank coating is used in ship drinking tank, freshwater tank and various fresh water cabinets. It should have good abilites of adhesion, water resistance and rust resistance, and paint film should be no water pollution and negative impact on human health. The selection of varieties need to be recognized and approved by the relevant health departments.

Drinking water tank coating usually includes three categories: lacquer phenol water tank coating, ketoimine curing epoxy coating and pure epoxy coating.

After the completion of the drinking water tank paint construction, the paint film should be completely solidified into the tank, so as to avoid some free harmful substances in the paint from infiltrating into the drinking water and affecting human health. After the coatings are completely cured, they will be generally soaked with fresh water from 2 to 3 days first. It can be put into use formally if the water quality qualifies through the water sample analysis.

2. Oil coatings

(1) The fuel oil tank generally is not required paint protection. In order to prevent the

corrosion of the bulkhead in the construction process and reduce the cleaning workload before sealing and refueling, a petroleum resin paint is often painted in the segmented painting stage. Petroleum resin paint is obtained by petroleum resin dissolved in hydrocarbon solvent and coated on the steel surface so that it can be dried into a film. When the fuel tank is filled with oil, the paint film will be dissolved gradually in the fuel. The bulkhead will contact directly with the fuel and will not be corroded.

(2) The sliding oil tank can be protected temporarily with petroleum resin paint as the oil tank, but the better way is to use pure epoxy coating protection, especially the main engine slide oil cycle tank. Its storage of oil quality requirements is higher, and it usually uses pure epoxy coating.

(3) The cargo tank of the crude oil ship is generally protected by epoxy asphalt coating. The coating parts are the bottom and roof, and the vertical bulkhead is usually not painted. If an inert gas protection system is provided in the cabin, the roof may not be painted.

(4) The cargo tanks of refined oil vessels and chemical vessels shall be fully painted. The use of the paint requirement is very high. It not only protects the inner surface of the oil tank from loading cargo and alternately ballast seawater corrosion, but also keeps the loaded goods away from pollution. It not only has strong resistance to cargo, but also has a certain heat resistance to adapt to the hot water washing tank.

New Words and Expressions

liquid ['lɪkwɪd] *n.* 液体
fresh water 饮用水
cooling water 冷却水
fuel oil tank 燃油舱
sliding oil tank 滑油舱
dirty oil and water tank 污油水舱
crude oil 原油
refined oil ship 成品油船
bow tip tank 艏尖舱
stern tip tank 艉尖舱
new shipbuilding 新造船
lacquer phenol water tank coating 漆酚水舱涂料
ketoimine [kiːtəʊˈɪmiːn] *n.* 酮亚胺
pure epoxy coating 纯环氧涂料
infiltrate [ˈɪnfɪltreɪt] *vt.& vi.* （使）渗透；（使）渗入
soak [səʊk] *vt.* 浸泡；浸透

bulkhead [ˈbʌlkhed] *n.* 舱壁
inert gas 惰性气体

Notes

1. The liquid compartment of ships is divided into two categories: water tank and oil tank. A water cabin includes fresh water tank, drinking water tank, ballast water tank, cooling water tank, bottom water tank, etc. The oil tank includes fuel oil tank, sliding oil tank, dirty oil and water tank, cargo oil tank of crude oil carrier, cargo oil tank of refined oil ship and chemical ship, etc.
Translation: 船舶的液舱分为水舱和油舱两大类。水舱包括淡水舱、饮水舱、压载水舱、冷却水舱、舱底水舱等。油舱包括燃油舱、滑油舱、污油水舱、原油运输船的货油舱、成品油船和化学品船的货油舱等。

2. In recent years, at the request of ship owners in the world, bleached epoxy asphalt coating without coal tar asphalt is used instead of traditional epoxy asphalt coating in the ballast tank of new shipbuilding.
Translation: 近年来，在世界各国船东的要求下，新造船的压载水舱内由采用传统的环氧沥青涂料改为采用不含煤焦沥青的漂白环氧沥青涂料。

Expanding Reading

According to the statistics of relevant departments, the number of man hours spent on painting operations of general ships accounts for about 12% to 15% of the total working hours of ship construction, and even more than 20% of special painting ships. In order to reduce the working time and intensity of rust removal painting operation, as well as reducing the harm of paint to human body and pollution to the environment, and improving the safety of the operation, the paint industry starts to improve the performance of the paint itself and clear the development direction of the current ship paint at present.

1. Shop primer with high heat resistance

Traditional inorganic zinc silicate shop primer can only withstand high temperature of 400℃, in order to reduce greatly the shop primer loss range of welding, cutting and fire correction time and reduce the workload of secondary rust removal. The new generation of shop primer can achieve a high temperature resistance from 800 to 1,000 ℃.

2. Rust-proof primer required for low surface treatment

At present, overseas primers with low surface treatment requirements have been introduced, including epoxy type, epoxy ester type, epoxy asphalt type and other primers. They are from the usual ISO Sa2-Sa2.5 and St3 to ISO St2 or even below, which can greatly reduce the workload of secondary rust removal.

3. Universal rust-proof primer

The same rust-proof paint may be used at any site of the vessel and any type of paint for paint and finish. It greatly reduces the workload of ship repair and painting surface treatment.

4. Thick film type or ultra-thick film type coatings

Thick film coatings or ultra-thick film coatings can fundamentally reduce the number of painting times. It has special significance for liquid tanks with complex structures and harsh working conditions. Many coating manufacturers have developed a film of more than 500 μm dry film thickness coatings, and even the thickness of coatings of up to more than 2,000 μm of ships and marine engineering corrosion prevention application of the epoxy coating.

5. Quick-drying type paint

It can shorten the curing, drying time of the coating and shorten the construction cycle. Especially in winter, two types (epoxy system and epoxy asphalt) of ship extensive application of paint are often difficult to dry and cure when the temperature is lower than 5 ℃, even after 2 to 3 days of coating surface is difficult to reach the degree of people walking on the surface. Fast-dry coatings can be reduced to 24 hours or less. There is a foreign fast dry epoxy asphalt coating that can achieve preliminary curing in about 8 hours under 5 ℃ condition.

6. Solvent-free or water-based rust-proof coatings

Solvent-free or water-based rust-proof coatings can reduce the air pollution of organic solvents and eliminate the risk of explosion and fire. It has special significance to the engine room with complex working environment, complex structures and the liquid chambers with difficult ventilation.

7. Low-toxic, non-toxic coatings

The development of low-toxic and non-toxic coatings can reduce the harm to human body to a certain extent, and there is great significance to protect the health of constructors and environment.

Exercises

I. Answer the following questions according to the passage.

1. What are the characteristics of liquid tank coatings?

2. Please describe how to choose the paint in the cargo holds of the tanker?

II. Practice these new words.

1. English to Chinese.

   fresh water _____   fuel oil tank _____

   refined oil ship _____   bulkhead _____

2. Chinese to English.

   原油 _____   艏尖舱 _____

   纯环氧涂料 _____   惰性气体 _____

III. Translation.

1. Translate the following sentences into Chinese.

   (1) After the completion of the drinking water tank paint construction, the paint film should be completely solidified into the tank, so as to avoid some free harmful substances in the paint from infiltrating into the drinking water and affecting human health.

   (2) The fuel oil tank generally is not required paint protection. In order to prevent the corrosion of the bulkhead in the construction process and reduce the cleaning workload before sealing and refueling, a petroleum resin paint is often painted in the segmented painting stage.

2. Translate the short passage.

   The ballast tank is used for various ballast water tank, bow tip tank, stern tip tank, cabin bottom water tank, etc. Due to the harsh corrosion environment of these tanks, the coatings are required to have excellent water resistance, salt mist resistance, dry and wet alternative resistance and corrosion resistance abilities. In addition, the construction conditions of these parts are harsh, so a higher film thickness is required for the coating to reduce the number of painting times.

Project 3 Pretreatment Before Painting

Background

The coating protection of ships is the most widely used, the oldest, the most economical, convenient and effective protection method. After the surface of steel is protected with coatings, its protection life is related to many factors. Table 3.1 shows the statistical analysis results of the influence of various factors on the coating life.

Table 3.1 Influence of Various Factors on the Coating Life

Influencing Factor	Influence Percentage/%
Surface Processing Quality	49.5
Film Thickness(the numbers of coatings)	19.1
Coating Types	4.9
Other Factors	26.5

Thus it can be seen, the quality of the steel surface treatment before the painting is the most important factor which can affect the performance of the coating. Therefore, the quality control of steel surface treatment is the most critical link to ensure the protection effect of the coating.

Steel surface treatment before painting is commonly known as rust removal, which not only refers to the removal of rust on the steel surface, but also should include the removal of the mill scale, old coatings and stained grease, dust, residual welding spatter and other pollutants. In addition, the steel after the surface treatment will also form a certain surface roughness. Therefore, the quality of steel surface treatment mainly refers to the degree of removal of the above pollutants, also known as "cleanliness", and the size of the roughness formed by the surface after the treatment.

Learning Target

Knowledge Target

1. Master the vocabulary related to painting pretreatment.
2. Master the pretreatment methods and tools before painting.

Ability Target

1. Translate between Chinese and English about surface treatment.
2. Proficient in using surface treatment related sentence patterns.

Task Definition

Task 3.1　Steel Pretreatment Assemble Line
Task 3.2　Spray Abrasive Surface Treatment
Task 3.3　Rust Removal by Chemicals
Task 3.4　Power Tool Cleaning

Task 3.1　Steel Pretreatment Assemble Line

Text Reading

In modern shipbuilding, there are 2 stages in ship painting steel surface treatment: Once steel gets into the factory, steel raw materials are removed the surface of the mill scale and rust and painting shop primer to ensure that steel does not continue to corrode in the processing. This stage of steel surface treatment is called steel surface pretreatment. The other stage of treatment is painting the steel surface treatment when the steel processing component section or then closed into the whole ship. It is usually called "secondary rust removal" in the shipbuilding industry. This project mainly introduces the surface treatment of shipbuilding steel. And the secondary rust removal method is similar to the surface treatment. Therefore, we do not mention it again.

Shipbuilding steel pretreatment method includes shot blasting treatment, blasting abrasive treatment and pickling treatment. At present, only shot blasting treatment can realize automatic assembly line operation with good quality and high efficiency.

The process flow of steel shot blasting pretreatment line is as follows:

1. Steel plate leveling

Steel plates for shipbuilding can be deformed during transportation or after a long period of stacking. The deformed steel plate will affect the processing accuracy, and the badly deformed steel plate will affect the line-type of the hull. Therefore, before or after the steel pretreatment, the steel plate should be leveled.

Steel plate leveling is usually used by seven star roller or nine star roller leveling machine. The leveling machine generally sets before the steel plate preheating treatment station, but some steel plate pretreatment assembly lines put the leveling machine behind the shot blasting machine, which is to protect the roller of the leveling machine to avoid the damage from the mill scale falling off the steel plate.

The ability of the leveling machine is different. The leveling machine of the shipbuilding steel plate is appropriate from 4 to 30 mm thick.

2. Steel plate transportation

After the steel plate feeding, the transmission of each process is completed by the roller channel. Rollers are usually cylindrical with bearing bases at both ends. The distance among roller channels is from 500 to 750 mm. In order to ensure the necessary processing time in the procedure, the pretreatment assembly line roller path must have sufficient quantity. Usually, if the head ramming ability is large and the steel plate conveying speed is fast, the number of roller channels should be more. Especially after the painting process, the necessary amount of rollers channel is provided to allow enough time for the shop primer to dry.

3. Preheating

Preheating is to heat up the steel plate before blasting and to remove the surface moisture and oil pollution, so that the steel plate can heat up to a certain temperature to facilitate the drying after painting. At present, the preheating equipment widely used at home and abroad is medium frequency induction heating, liquefied petroleum gas heating and hot water spray heating. Regardless of the preheating method used steel plate should be heated up to about 40 ℃. If the temperature is too low, it will not be conducive to the removal of water and oil or the subsequent spraying shop primer drying. And if the temperature is too high, in addition to increasing energy consumption, it will be easy to make the shop primer because of too fast drying and foaming.

4. Shot blasting

The shot blasting is done in the shot blasting room. The shot blasting room is equipped with shot blasting device (commonly known as head ramming), abrasive circulation device, abrasive cleaning device, ventilation and dust removal device, etc.

The shot blasting device is composed of impeller, shield, directional sleeve, ball wheel, bearing seat and motor. The impeller is rotated at high speed (from 2,200 to 2,600 r/min) to produce a strong centrifugal force. When the abrasive is sucked into the ball wheel through the ball tube, the movement is accelerated along the blade length under the centrifugal force, then throw it until a speed from 60 to 80 m/s. The abrasive thrown out forms a fan-shaped flow beam and strikes the surface of steel plate to remove mill scale and rust.

5. Spraying

The steel surface shall be coated with the shop primer immediately after shot blasting. That is an automatic progress. The whole painting device is composed of high-pressure airless spray machine, automatic spray gun, ventilation and fog removal device, etc. The automatic spray gun is moved by cylinder or chain and used by the stroke switch to control the start and stop. It is placed on both sides that up and down of the steel plate, running in the opposite direction. The distance between the spray gun and the steel plate is usually about 300 mm, and the lower gun should be slightly less than the upper gun.

Due to prevent pollution, air pipes, paint filters, fans and exhaust pipes shall be installed in the spray room, the exhaust volume depends on the solvent volatilization of the shop primer and the allowable solvent gas concentration.

6. Dry

The steel plate should enter the drying furnace to promote rapid drying and facilitate rapid transfer after painting. Drying furnace can be far infrared radiation or steam as heat source, but it can not be heated directly by open fire. The drying furnace shall be equipped with exhaust device to prevent the solvent gas accumulation in the furnace and causing explosion accidents.

For good dry shop primer (such as inorganic zinc silicate shop primer), generally, the drying process can be exempted.

New Words and Expressions

mill scale [ˌɒksɪˈdeɪʃn] *n.* 氧化皮
shipbuilding industry 造船工业
shot blasting 抛射磨料处理
blasting treatment 喷射磨料处理
pickling treatment 酸洗处理
leveling [ˈlevəlɪŋ] *n.* 校平
transportation [ˌtrænspɔːˈteɪʃn] *n.* 运输
stacking [ˈstækɪŋ] *n.* 堆放；堆积
accuracy [ˈækjərəsi] *n.* 精确（性）；精度
fall off 脱落；跌落
roller channel 辊道
sufficient [səˈfɪʃnt] *adj.* 足够的；充足的
preheat [ˌpriːˈhiːt] *n.* 预热
medium frequency induction heating 中频感应加热
liquefied petroleum gas heating 液化石油气加热
hot water spray heating 热水喷淋加热
energy [ˈenədʒɪ] *n.* 能量
consumption [kənˈsʌmpʃn] *n.* 消耗
foam [fəʊm] *vi.* 起泡
device [dɪˈvaɪs] *n.* 装置
abrasive [əˈbreɪsɪv] *n.* 磨料
impeller [ɪmˈpelə] *n.* 叶轮；推进器
centrifugal force 离心力
spray gun 喷枪
cylinder [ˈsɪlɪndə(r)] *n.* 气缸
filter [ˈfɪltə(r)] *n.* 过滤器

fan [fæn] *n.* 风机
exhaust pipes 排气管道
concentration [ˌkɒnsnˈtreɪʃn] *n.* 浓度
drying furnace 烘干炉
far infrared radiation 远红外线辐射
steam [stiːm] *n.* 水蒸气
inorganic zinc silicate 无机硅酸锌

Notes

1. Shipbuilding steel pretreatment method includes shot blasting treatment, blasting abrasive treatment and pickling treatment.
   Translation: 造船用钢材预处理的方式有抛射磨料处理、喷射磨料处理和酸洗处理三种方式。
2. The shot blasting device is composed of impeller, shield, directional sleeve, ball wheel, bearing seat and motor.
   Translation: 抛丸器由叶轮、护罩、定向套、分丸轮、轴承座及电动机等组成。

Expanding Reading

The control of the steel surface treatment quality before painting mainly includes two aspects: the cleanliness and roughness of the steel surface.

The national standard GB 8923 is equivalent to the international standard ISO 8501-1:1998. This standard divides the original rust degree of the uncoated steel surface into four "rust grades", and the unpainted steel surface and the quality of the fully removed original coated steel surface are divided into several "rust removal grades".

1. Rust grade (Figure 3.1)

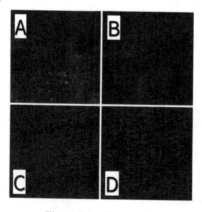

Figure 3.1 Rust grade

A. Steel surface is fully covered with mill scale and almost no rust;

B. Steel surface has been corroded and some mill scale has been flake;

C. Steel surface that has been flaked by rust, or can be scraped, and has a small amount of pitting;

D. Steel surface that has been completely stripped by rust and commonly eroded.

2. Rust removal grade (Figure 3.2)

Figure 3.2   Rust removal grade by spraying

(1) Spray or blast to remove rust

Sa1 Mild jet or blast rust

The surface of the steel shall be free of visible grease and dirt, and no weak adhere such as mill scale, rust and old coatings.

Sa2 Thoroughly spray or blast rust removal

The surface of the steel shall be free of visible grease and dirt, and the adherements such as mill scale, rust and old coatings have been basically removed, and its residues shall be firmly attached.

Sa2.5 Very thorough spray or blast rust removal

The surface of steel shall be free from visible grease, dirt, mill scale, rust, and old coatings, and any residual trace shall be only dotted or striped slight color spots.

Sa3 Make the steel apparent clean, and spray or blast for rust removal

The steel surface shall be free of visible grease, dirt, mill scale, rust, and old coatings. The surface shall show a uniform metallic color luster.

(2) The manual and power tool for rust removal (Figure 3.3)

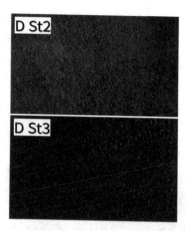

Figure 3.3 Rust removal grade by power tool

St2 Thorough manual and powered tools for rust removal

The surface of the steel shall be free of visible grease and dirt, no weak adhere mill scale, rust and old coatings.

St3 Very thoroughly manual and power tool for rust removal

The surface of the steel shall be free of visible grease and dirt, and no weak adhere mill scale, rust and old coatings. The rust removal shall be more thorough than St2 and the surface of the exposed substrate shall be metallic luster.

(3) Flame rust removal

The steel surface shall be free of mill scale, rust and old coatings, and any residual trace is only the surface discoloration (shadows of different colors).

Exercises

I. Answer the following questions according to the passage.

1. How many steps are there in the steel pretreatment assembly line? What are they?

2. Why do you want to preheat the steel plate when conducting the pretreatment?

II. Practice these new words.

1. English to Chinese.

   mill scale _____   preheat _____
   foam _____   abrasive _____

2. Chinese to English.

抛射磨料处理 _____   酸洗处理 _____

喷枪 _____   过滤器 _____

III. Translation.

1. Translate the following sentences into Chinese.

(1) The leveling machine generally sets before the steel plate preheating treatment station, but some steel plate pretreatment assembly lines put the leveling machine behind the shot blasting machine, which is to protect the roller of the leveling machine to avoid the damage from the mill scale falling off the steel plate.

(2) The whole painting device is composed of high-pressure airless spray machine, automatic spray gun, ventilation and fog removal device, etc.

2. Translate the short passage.

In modern shipbuilding, there are 2 stages in ship painting steel surface treatment: Once steel gets into the factory, steel raw materials are removed the surface of the mill scale and rust and painting shop primer to ensure that steel does not continue to corrode in the processing. This stage of steel surface treatment is called steel surface pretreatment. The other stage of treatment is painting the steel surface treatment when the steel processing component section or then closed into the whole ship. It is usually called "secondary rust removal" in the shipbuilding industry.

Task 3.2 Spray Abrasive Surface Treatment

Text Reading

Shot blasting (sand) treatment is a high efficiency surface treatment method that sprays the abrasive to the treated steel surface at a certain speed, and removes the mill scale, rust and pollution on the steel surface with the abrasive of impact and grinding action.

Compared with manual cleaning and power tool cleaning, shot blasting treatment is much higher cleaning efficiency and quality. It is also less investment, less footprint, fast operation, strong motility and simple operation and maintenance compared with shot cleaning. And compared with pickling, it is the treatment that is low environmental pollution and easy to control and no damage of raw material performance. In addition, large surface roughness for improving coating adhesion is an outstanding advantage.

Main system description:

1. Sand blasting system

Sand blasting system generally adopts the most advanced two-cylinder four (two) gun continuous sandblasting machine, or chooses single-cylinder single (double) gun sandblasting machine.

2. Floor and sand return system

It is pushed to the pit by the sand truck, then sent to the bucket elevator by the belt conveyor, and enters the dust pellet separator. The screened and separated abrasive enters the sand storage box. Then the abrasive adopts self-flow distribution, and sets up manual material valve and automatic material valve.

3. Local dust removal system

After the steel sand bucket elevator, it earns the dust shot separator. At this time, a large amount of dust will be extracted through the dust removal machine, and local dust removal adopts secondary dust removal. The first stage is cyclone dust removal which can remove 70% of the dust, and the second stage is the filter tube dust removal. The air will be discharged after treatment meets the national environmental protection emission standards.

4. All-room ventilation and dust removal system

Sand blasting operation will produce a large amount of dust, and must be ventilated and removed dust in time. The dust removal and ventilation capacity shall be calculated from 10 to 20 times per hour. In order to ensure good humidity conditions in the sandblasting room and save energy, the treated wind will be returned into the sandblasting room. The filtration requirement is a dust removal capacity of 99.999% for 0.5 μm dust. The air will be discharged after treatment meets the national environmental protection emission standards.

5. Block sand return system

If it is the sand blasting for block, considering the investment and operation cost, most of the steel sand left in the block is recycled manually, and the bottom residue is recycled by vacuum cleaner.

6. Dehumidification system

Dehumidification system generally adopts rotary wheel and frozen heating dehumidifier, which is suitable for all-weather operation. At the same time, its computer control system will automatically consider energy saving and convenience, and open different dehumidification modes.

7. Compressed air drying system

The quality of compressed air determines the quality of sandblasting treatment. In principle, the particles shall be less than 1 μm, oil content is less than 1 mg/m^3 and pressure dew point temperature -20 ℃. The compressed air into the sandblasting room first enters the atmospheric bag, then enters the pre-filter and the cold dryer, and finally outputs through the filter.

8. Electrical control system

In principle, the centralized control room system consists of PLC. Power distribution cabinet and control box are set in each machine room. The equipment with electric control box shall be controlled by site, but all operation, fault and other signals are fed to the centralized control room.

New Words and Expressions

power tool cleaning 动力工具清理
sand blasting system 喷砂系统
machine [məˈʃiːn] *n.* 机器，机械
floor and sand return system 地坪回砂系统
sweep [swiːp] *v.* 打扫
pit [pɪt] *n.* 凹陷，地坑
belt [belt] *n.* 腰带；传送带
conveyor [kənˈveɪə(r)] *n.* 运送者，传送者
bucket elevator 斗式提升机
separator [ˈsepəreɪtə(r)] *n.* 分离器，分离装置
extract [ˈekstrækt] *v.* 抽出；提取
cyclone [ˈsaɪkləʊn] *n.* 气旋，旋风
filter tube 滤管，滤筒
emission [iˈmɪʃn] *n.* 排放
filtration [fɪlˈtreɪʃn] *n.* 过滤

rotary ['rəʊtərɪ] *adj.* 旋转的

frozen ['frəʊzn] *adj.* 冰冻的，冰封的

compressed air 压缩空气

pre-filter 预过滤

centralized control room 集控室

Notes

1. Shot blasting (sand) treatment is a high efficiency surface treatment method that sprays the abrasive to the treated steel surface at a certain speed, and removes the mill scale, rust and pollution on the steel surface with the abrasive of impact and grinding action.

   Translation: 喷丸（砂）处理是以压缩空气为动力，将磨料以一定的速度喷向被处理的钢材表面，以磨料对钢材表面的冲击和磨削作用，将钢材表面的氧化皮、锈蚀产物及其他污物除去的一种高效率的表面处理方法。

2. Sand blasting operation will produce a large amount of dust, and must be ventilated and removed dust in time.

   Translation: 喷砂作业时会产生大量粉尘，必须及时进行通风除尘。

Expanding Reading

Shot blasting (sand) treatment is very complicated work, therefore, the operation should pay attention to the following matters.

1. Preparation work

(1) Remove the debris, garbage and seeper in the block;

(2) Check the block situation, whether it is smooth or not. It affects the entry and exit of personnel;

(3) Build up the scaffolding according to the operation needs;

(4) Remove the seeper and dirty oil in the gas storage tank and the oil-water separator;

(5) Fill the shot blasting cylinder with abrasive;

(6) Connect the shot blasting hose and spray tube, then check whether the hose is damaged and the joint is firm or not. Make the spray gun in place;

(7) Connect the low-voltage portable lighting fixtures;

(8) Wear good personal protective equipment;

(9) Close the door of the shot blasting room.

2. Operating requirements

(1) Open the all-room ventilator to make the indoor all-room ventilate;

(2) Open the belt conveyor and the bucket elevator (the pneumatic recovery system can be

omitted);

(3) Open the shot blasting cylinder pressure valve, then open the enter air valve, and then open the outer air valve and use the outer shot volume and inlet air volume to reach a better mixing ratio;

(4) The shot blasting operation should be carried out in blocks, first down and up, from down to up, inside to outside and difficult to easy. The spraying gun should be from 70 to 85 degrees to the treated surface, and the distance between the nozzle and the surface should be about 300 mm;

(5) After the shot blasting, close the outer shot valve, and then close the enter shot valve after the abrasive is exhausted in the hose. Finally open the exhaust valve to remove the remaining gas in the cylinder;

(6) Eliminate the abrasive accumulated in the block, suck the abrasive into the volume separator with a vacuum recycling machine, and then discharge the iron ball in the volume separator into the shot pit;

(7) Open the powered recovery device and collect the abrasive in the ball pit (the mechanical recovery device can be omitted);

(8) Close the abrasive recovery device (the mechanical recovery device shall first close the belt conveyor, and then close the bucket elevator);

(9) Shut down the full-room ventilator;

(10) Quality inspection and defect repair;

(11) Remove the shot blasting pipe and low-voltage lighting, and open the door of the shot blasting room.

Exercises

Ⅰ. **Answer the following questions according to the passage.**

1. Can you describe characteristics of shot blasting treatment in your own words?

2. Can you briefly introduce the composition of the shot blasting system?

Ⅱ. **Practice these new words.**

1. English to Chinese.

   sand blasting system _____   bucket elevator _____

   separator _____   filter tube _____

2. Chinese to English.

机器，机械 _____   地坪回砂系统 _____

过滤 _____   压缩空气 _____

III. Translation.

1. Translate the following sentences into Chinese.

(1) In order to ensure good humidity conditions in the sandblasting room and save energy, the treated wind will be returned into the sandblasting room.

(2) If it is the sand blasting for block, considering the investment and operation cost, most of the steel sand left in the block is recycled manually, and the bottom residue is recycled by vacuum cleaner.

2. Translate the short passage.

Compared with manual cleaning and power tool cleaning, shot blasting treatment is much higher cleaning efficiency and quality. It is also less investment, less footprint, fast operation, strong motility and simple operation and maintenance compared with shot cleaning. And compared with pickling, it is the treatment that is low environmental pollution and easy to control and no damage of raw material performance. In addition, large surface roughness for improving coating adhesion is an outstanding advantage.

Task 3.3 Rust Removal by Chemicals

Text Reading

For the pretreatment of steel raw materials, pickling treatment is often used in small and medium-sized yards without shot blasting pretreatment assembly line, but also for some large yards in the treatment of sheet and pipe application. Generally, rust removal of the outfitting parts and components is mainly used in acid pickling treatment.

Apickling is a chemical reaction of inorganic acid or organic acid with steel surface mill scale (molecular formula Fe_3O_4, actually Fe_2O_3 and FeO compound), rust (composite hydrate of iron oxide and ferrous oxide) to form soluble iron salt, and process means that removes it from the steel surface.

Inorganic acids and organic acids are often used for pickling. Inorganic acids include sulfuric acid, hydrochloric acid, phosphoric acid, nitric acid, hydrofluoric acid, etc. Organic acids include citric acid, gluconic acid, low-carbon delspray, etc.

There are several advantages for inorganic acids, such as strong forcing, fast rust removal speed, wide raw material source and low price, and the disadvantage is easy to produce overcorrosion phenomenon, even if the corrosion inhibitor is often difficult to escape. And if cleaning is not thorough, residual acid will continue to corrode steel, so that the steel surface coating blisters or even falls off.

The speed of organic acid pickling is slow. There is no serious future problem about the residual acid. It is not easy to rust again, and the buffer solution is formed in the pickling process which is easy to control the pH value. After the treatment, the substrate surface is clean, and the use period of the pickling liquid is longer. However, organic acids are much more expensive than inorganic acids and have less chemical force. So they are mostly used for power container internal cleaning and other pickling occasions with specific requirements.

The process of pickling is roughly as follows:

1. Removing oil

Alkaline emulsifier is used. The temperature is generally from 70 to 80℃, and the time is about 1 hour (according to the degree of greasy can be appropriately shortened or extended). However, when the surface is covered with a large number of oil stains and impurities, it should be scraped off in advance.

2. Washing water

Immersion and spray combination are used. It is to clean the defat solution on the inner and outer surface of the workpiece.

3. Acid treatment

After the washing workpieces, they are immersed immediately in the acid tank, acid tank with acid, corrosion inhibitor, wetting agent, etc. The pickling temperature is generally from 40 to 60 ℃, and the dipping time is controlled according to the corrosion of the workpieces to avoid excessive corrosion caused by too long time. With the working time of acid in the acid tank, the acidity will be decreased, and the inhibitor will be consumed. It should be supplemented in time. Mixing is conducive to improving the rust removal speed and reducing the acid consumption.

4. Wash again

The purpose is to clean the acid solution on the surface of the workpieces. The methods are the same to the second step. When washing, the workpieces should be moved up and down to improve the cleaning effect.

5. Neutralization processing

Workpieces which are washed with cold water still have a small amount of dilute acid on the surface and should be neutralized in dilute alkali.

6. Wash it with cold water

After neutralization, the workpieces must also be rinsed again with cold water to remove the small amount of salt formed from the neutralization solution and the surface.

7. Phosphorization treatment

After pickling, the steel surface is activated and easy to corrode in the air. Usually it should be made for phosphating treatment, so that the steel surface forms a water-insoluble metal (iron) phosphate film for protecting the steel plate from corrosion in a short of time.

8. Soap

Steel plate with cold deformation usually needs to add a saponification process after the phosphating treatment. Saponification is generally performed with stearate, the temperature is controlled from 80 to 90 ℃.

After the pickling liquid is used to a certain extent, the concentration of iron ions in the solution is greatly increased, and the cleaning ability is greatly reduced. So it cannot continue to be applied, and it should be replaced by the new acid liquid. Therefore, the waste acid can be treated in the following two methods: One is to absorb residual acid from iron liquid, making ferrous salts, or alkali neutralization; The other is to use ion exchange resin method, electrodialysis method, and oxidation precipitation method to remove waste acid iron ions, so that the waste acid become new.

New Words and Expressions

raw [rɔː] *adj.* 生的；自然状态的
pickling [ˈpɪklɪŋ] *n.* 酸洗

yard [jɑːd] *n.* 船厂；场地
sheet [ʃiːt] *n.* 薄板
pipe [paɪp] *n.* 管道；管材
inorganic acid 无机酸
organic acid 有机酸
mill scale 氧化皮
compound [ˈkɒmpaʊnd] *n.* 化合物，混合物
ferrous oxide 氧化亚铁
overcorrosion [ˌəʊvərkəˈrəʊʒ(ə)n] *n.* 过腐蚀
emulsifier [ɪˈmʌlsɪfaɪə(r)] *n.* 乳化剂
impurity [ɪmˈpjʊərəti] *n.* 污点，污染
scrape [skreɪp] *vt.* 刮；擦
defat [diːˈfæt] *vt.* 使……脱脂
workpiece [ˈwɜːkˌpiːs] *n.* 工件
inhibitor [ɪnˈhɪbɪtə(r)] *n.* 抑制剂
neutralize [ˈnjuːtrəlaɪz] *n.* 中和
phosphorization [fɒsfəraɪˈzeɪʃən] *n.* 磷化作用
soap [səʊp] *n.* 皂化

Notes

1. Apickling is a chemical reaction of inorganic acid or organic acid with steel surface mill scale (molecular formula Fe_3O_4, actually Fe_2O_3 and FeO compound), rust (composite hydrate of iron oxide and ferrous oxide) to form soluble iron salt, and process means that removes it from the steel surface.
   Translation: 酸洗是应用无机酸或有机酸与钢铁表面的氧化皮（其分子式为Fe_3O_4，实际上是Fe_2O_3与FeO复合化合物）、铁锈（氧化铁和氧化亚铁的复合水合物）进行化学反应，生成可溶性铁盐，然后将其从钢铁表面清除的工艺手段。

2. The pickling temperature is generally from 40 to 60 ℃, and the dipping time is controlled according to the corrosion of the workpieces to avoid excessive corrosion caused by too long time.
   Translation: 酸洗温度一般在40~60 ℃，浸渍时间根据工件锈蚀情况控制，避免时间过长而引起过蚀现象。

Expanding Reading

Phosphorization of steel refers to the process of generating various water-insoluble metal phosphate coatings on the surface of iron and steel through chemical treatment.

The main purpose of phosphorization is to improve the corrosion resistance of the steel surface, which can be used as the bottom layer of the coating protection. It is to improve the adhesion between the steel and the organic coating, also to improve the wear resistance of the sliding surface.

There are many ways of phosphorization:

(1) According to the classification of the treatment solution components: zinc phosphate line, calcium zinc phosphate line, manganese phosphate line, and iron phosphate line;

(2) According to the treatment temperature classification: high temperature phosphorization (from 90 to 98 ℃), medium temperature phosphorization (from 50 to 70 ℃), and low temperature phosphorization (from 20 to 30 ℃);

(3) According to the treatment process classification: tank immersion, spray, and brush phosphorization.

Phosphating in different ways can obtain phosphating films with different thickness and compositions to meet different needs.

Exercises

I. Answer the following questions according to the passage.

1. What are the inorganic acids that can be used for pickling?

2. How can we avoid the overerosion phenomenon?

II. Practice these new words.

1. English to Chinese.
   pickling _____   pipe _____
   organic acid _____   inhibitor _____
2. Chinese to English.
   无机酸 _____   氧化皮 _____
   过腐蚀 _____   乳化剂 _____

III. Translation.

1. Translate the following sentences into Chinese.

(1) There are several advantages for inorganic acids, such as strong forcing, fast rust removal speed, wide raw material source and low price, and the disadvantage is easy to produce

overcorrosion phenomenon, even if the corrosion inhibitor is often difficult to escape. And if cleaning is not thorough, residual acid will continue to corrode steel, so that the steel surface coating blisters or even falls off.

(2) The speed of organic acid pickling is slow. There is no serious future problem about the residual acid. It is not easy to rust again, and the buffer solution is formed in the pickling process which is easy to control the pH value. After the treatment, the substrate surface is clean, and the use period of the pickling liquid is longer.

2. Translate the short passage.

Therefore, the waste acid can be treated in the following two methods: One is to absorb residual acid from iron liquid, making ferrous salts, or alkali neutralization; The other is to use ion exchange resin method, electrodialysis method, and oxidation precipitation method to remove waste acid iron ions, so that the waste acid become new.

Task 3.4 Power Tool Cleaning

Text Reading

Power tool grinding treatment refers to the use of a variety of wind or electric rust removal tools and relies on the power motor to high speed rotation or reciprocating movement to drive grinding equipment (grinding wheel, sandpaper disk, wire brush, chipping hammer, etc.) to grind or strike. The surface which needs to be painted on, in conclusion, is a mechanical cleaning way to remove rust and other debris (Figure 3.4).

Figure 3.4 Commonly tool of power tool cleaning

Power tools are small in volume, light in weight and easy for individuals to carry and operate, so they have strong adaptability. Compared with quality of the old manual tool to knock out the rust, the power tool grinding treatment has the advantages of good rust removal quality and high production efficiency. However, this kind of rust removal method belongs to the semi-mechanized operation of manual operation, which is low inefficient, poor quality and small surface roughness, so it is limited. Especially for some special high performance coatings, power tool grinding treatment is not suitable for the quality requirements of coatings on surface treatment.

Commonly used power tools are straight handle grinding turbine, end-type flat sand grinding machine, flat sand grinding machine with cone gear and chipping hammer.

Key points of power tool grinding process:

(1) Observe the working environment, scaffolding, lighting and other auxiliary work. Prepare grinding tools, air duct joints and personal labor protection supplies;

(2) Clean up the surrounding environment (including garbage, debris and water in the block);

(3) Install the sandpaper disk on the end-type flat sand mill or the flat sand mill with cone gear, and thoroughly polish the weld area, fire burning area and natural rust area until it shows the

metal color. The grinding operation should follow principles of "difficult then easy" and "down before up";

(4) Gently polish the intact parts of the shop primer to remove the surface zinc salt (for the zinc-containing shop primer) or the surface aging layer (for the shop primer without zinc);

(5) For the secondary rust removal in the area, the surrounding coating on both sides of the weld and the burning area and the natural rust area should be made into a slope to help repair the adhere to the coating superposition;

(6) Change the bowl-shaped or dish-shaped steel wire brush plate, and thoroughly brush the weld, burning area and natural rust area, so that the yellow rust in the fine holes can be removed;

(7) Install the bundle (pen) steel wire brush with straight handle grinding turbine (small specification), and brush the corner joint around the hole and the tools are difficult to reach to remove yellow rust;

(8) Use a chipping hammer to remove the scattered iron scales and a small amount of the weld slag;

(9) Clean up garbage and dust accumulation;

(10) Remove oil stains on the surface with broken steps or gauze balls dipped in solvent;

(11) Submit it for quality acceptance.

New Words and Expressions

reciprocating [rɪˈsɪprəkeɪtɪŋ] *adj.* 往复的；来回的；交替的
grinding wheel 砂轮
sandpaper plate 砂纸盘
wire brush 钢丝刷
chipping hammer 风铲；气铲
strike [straɪk] *v.* 碰撞；打击
individual [ˌɪndɪˈvɪdʒuəl] *adj.* 单独的；个人的
efficiency [ɪˈfɪʃnsɪ] *n.* 效率
straight handle grinding turbine 直柄砂轮机
end-type flat sand grinding machine 端型平面砂磨机
cone [kəʊn] *n.* 圆锥体
gear [ɡɪə(r)] *n.* 齿轮
scaffolding [ˈskæfəldɪŋ] *n.* 脚手架
debri [ˈdebriː] *n.* 碎片；垃圾
mill [mɪl] *vt.* 研磨；粉碎；*n.* 砂磨机
intact [ɪnˈtækt] *adj.* 完整无缺的
superposition [ˌsjuːpəpəˈzɪʃn] *n.* 重叠；叠加

scattered [ˈskætəd] *adj.* 零散的；分散的
dip [dɪp] *v.* 沾；蘸
gauze [gɔːz] *n.* 纱布

> **Notes**

1. Power tool grinding treatment refers to the use of a variety of wind or electric rust removal tools and relies on the power motor to high speed rotation or reciprocating movement to drive grinding equipment (grinding wheel, sandpaper disk, wire brush, chipping hammer, etc.) to grind or strike. The surface which needs to be painted on, in conclusion, is a mechanical cleaning way to remove rust and other debris.
Translation: 动力工具打磨处理，是指采用各种风动或电动的除锈工具，依靠动力马达高速旋转或往复运动带动打磨器具（砂轮、砂纸盘、钢丝刷、气铲等）磨削或打击需要涂装的表面，达到清除铁锈及其他杂物的一种机械清理方式。

2. Power tools are small in volume, light in weight and easy for individuals to carry and operate, so they have strong adaptabiliy.
Translation: 动力工具体积较小，质量较小，便于个人携带和操作，应变能力较强。

> **Expanding Reading**

After the pre-treated steel is blocked, some of the shop primers are damaged of the steel surface due to welding, cutting, mechanical collision or natural reasons, resulting in re-corrosion of the steel surface. After the block closure, in the regional painting stage, there are always some block coatings, which are damaged and seriously rusted due to the above reasons. This requires another surface treatment. Compared to the raw material pretreatment (the first rust removal), this is the second surface treatment, so it is called the "secondary rust removal".

In addition to blasting abrasive treatment and power tool grinding treatment, there are many ways of secondary rust removal.

1. Vacuum shot blasting (sand) treatment

Vacuum shot blasting (sand) treatment is to use compressed air injection to vacuum room air extraction, so that the suction pipe will connect the vacuum chamber to the cone cover and form the negative pressure difference. So as to spray the abrasive and the rust removed into the vacuum device. Its biggest advantage is that it does not pollute the environment, especially in the ship outfitting stage. The general sandblasting treatment should not only affect other types of work, but also bring damage to the installed machinery and instruments. There will be no these disadvantages, if the vacuum is used. However, the vacuum pellet (sand) treatment efficiency is low, and it is difficult to recover abrasives and dust for uneven surfaces and corners.

2. Wet sandblasting treatment

Wet sandblasting treatment is to add a part of water to abrasive injection to the treated surface. It can effectively r' etc. Its advantage is that it can greatly reduce the dust fly beneficial to environmental protection and the health of op

After wet sandblasting, the surface appears in a wet state, with phenomenon. So a certain amount of corrosion inhibitor (about 1% of the w. be added in the water.

3. Water (abrasive) blasting treatment

Water (abrasive) blasting treatment is to use the water mixed with abrasive (sand), under the 0.6 to 0.8 MPa air pressure blasting to the treated surface. It can effectively remove the mill scale, rust, old coatings, etc. Its advantages are not only can greatly reduce the dust flying in the sandblasting site, but also can avoid the abrasive and metal strike sparks, which are conducive to safe operation. In addition, the water-soluble salt on the treated surface can be removed simultaneously. During repair ship surface treatment, layered quality degradation and adhesion reduced coating can be removed.

After water (abrasive) blasting treatment, the surface will be contaminated by wet mortar, and should be washed immediately with fresh water while it is still wet. The water used for washing undoubtedly needs to be added an inhibitor which is used in about 5% of the amount of water. After washing, the surface should be managed to dry and painted as soon as possible.

Exercises

I. Answer the following questions according to the passage.

1. Use your own language to describe the advantages and disadvantages of power tool cleaning.

2. What are the operational points of power tool cleaning?

II. Practice these new words.

1. English to Chinese.

   sandpaper plate _____   grinding wheel _____

   wire brush _____   scaffolding _____

2. Chinese to English.

气铲 _____ 砂轮机 _____

齿轮 _____ 研磨 _____

III. Translation.

1. Translate the following sentences into Chinese.

(1) Install the sandpaper disk on the end-type flat sand mill or the flat sand mill with cone gear, and thoroughly polish the weld area, fire burning area and natural rust area until it shows the metal color. The grinding operation should follow principles of "difficult then easy" and "down before up".

(2) For the secondary rust removal in the area, the surrounding coating on both sides of the weld and the burning area and the natural rust area should be made into a slope to help repair the adhere to the coating superposition.

2. Translate the short passage.

Power tools are small in volume, light in weight and easy for individuals to carry and operate, so they have strong adaptabiliy. Compared with quality of the old manual tool to knock out the rust, the power tool grinding treatment has the advantages of good rust removal quality and high production efficiency. However, this kind of rust removal method belongs to the semi-mechanized operation of manual operation, which is low inefficient, poor quality and small surface roughness, so it is limited. Especially for some special high performance coatings, power tool grinding treatment is not suitable for the quality requirements of coatings on surface treatment.

Project 4 Ship Painting

Background

Due to the huge surface area, complex structures, a wide variety of painting, long construction cycle and outdoor operation of ships, there are many differences in ship painting operation and the general steel painting of manufacturing products.

In order to obtain good quality painting and protect the ship effectively, in addition to doing serious surface treatments and secondary rust removal for the hull steel surface, choosing good quality painting and reasonable matching is also important. We need to use correct coatings, ensure a good working environment and adopt appropriate operation methods with scientific phased operation.

Learning Target

Knowledge Target

1. Understand the words and sentence patterns related to the preparation work before painting.

2. Master the vocabulary and their usage related to the painting operation methods.

3. Master the characteristics of brush painting, roller painting, compressed air spraying and high-pressure airless spraying and other related words and usage.

Ability Target

1. Skillfully use painting equipment, painting process and other related words.

2. Be able to read the English painting specifications, and grasp the general idea.

Task Definition

Task 4.1   Preparation Before Painting
Task 4.2   Brush Painting and Roller Painting
Task 4.3   Compressed Air Spraying
Task 4.4   High-Pressure Airless Spraying

Task 4.1 Preparation Before Painting

Text Reading

The ship painting operation workload is large, and the working environment is complex and changeable, thus we must be fully prepared before painting.

1. Open the pot

Before opening the paint pots, it is necessary to check carefully and confirm whether the variety, brand, color, and factory date meet the specified requirements. If the above marks on the paint pots are blurred, the relevant warehouse records should be carefully checked for confirmation before the pots are opened. To the coatings that the factory date exceeds the regulation, we should identify carefully whether there is gelling, block and other malpractice after opening the pots, and use it only after confirming that there is no obvious change in its performance.

After the paint is opened, if there is crust on the surface, the crust should be removed along the edge of the pot and should not be mashed into the paint. The paint details are shown in Figure 4.1.

Figure 4.1 Paint details

2. Stir

During the storage period of coatings, the phenomenon of pigment settles at the bottom of the pot because of its large density, while some pigments will have "segregation" floating on the surface due to their poor dispersion. Therefore, it must be carefully stirred before using. When stirring, turn up the pigment deposited at the bottom of the pot with smooth and clean batten, and then stir with an electric or pneumatic mixer until the viscosity of the coatings is uniform.

If the pigment sink is serious, hardening and can not be stirred, we can use the upper coatings, and dump to the bottom of the plate, or the performance of the coatings will change, and the quality can not be guaranteed. For such paint, if it occurs in the storage validity period, it should be returned to the paint factory for replacement. Similarly, if there is gelling during the storage period of the coatings, it should also be replaced by the paint factory.

The harden and gelling coatings cannot be forcibly diluted and shall be scrapped.

3. Mix and Cure

The coating that cured by chemical reaction generally is a two-component type (even a three-component type), and the binder and curing agent of two-component coatings are packaged respectively, and mixed when in use. Once mixed, it needs to be used up in the prescribed time. Beyond this prescribed time, the paint will lose fluidity, or even cure. Therefore, according to the need to be mixed by the prescribed proportion, so as not to cause waste.

Before mixing, the binder and curing agent should be mixed evenly in advance, and then mixed while mixing. Finally, they are all uniform.

Some two-component coatings require a period of pre-reaction before use, also known as curing. This is to ensure its construction performance and curing performance. The specified curing time should be guaranteed according to the requirements of the coatings product specification.

4. Dilute

The viscosity of the coatings has been adjusted in the factory. It can be used after being opened and stirred evenly. However, if the coatings are thickened for storage reasons, or due to construction that the viscosity needs to be reduced, it should be diluted with thinner.

Each coating has a corresponding thinner and a specified maximum dilution, which shall be subject to the provisions of the coatings product specification. Using the wrong thinner or excessive dilution will reduce the construction performance and internal quality of the coating, and even cause scrap. After the thinner is added into the coatings, it should be fully stirred until the whole is uniform.

5. Filter

Paint in the storage process may produce a skinning. Its pigment particles will also gather together to form a larger particles, although after full mixing, it is often difficult to disperse. This will cause the nozzle blockage when spraying, and will also affect the smooth appearance of the coatings. Therefore, it is necessary to filter the coatings after mixing (including mixing after dilution).

Filtering generally uses from 60 to 80 μm filter (Figure 4.2). The filter screen should be clean and cleaned with solvent immediately after the use. If the filtered paint is not used temporarily, it should be covered for protection to prevent impurities from mixing.

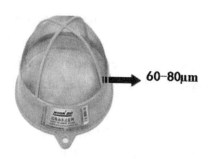

Figure 4.2 Filter from 60 to 80 μm

6. Color

The finish paint color of the new shipbuilding is usually prepared by the paint factory according to the requirements of the shipyard. But sometimes, especially the amount of less, it is used as a variety of logo paint, not necessary everything to the paint factory customization, and it needs to be prepared by itself.

The color of the paint should be mixed with the same binder, according to the proportion of different colors. The most basic colors are red, yellow and blue. Any two or three of these three primary colors can be mixed in different proportions to obtain other colors. Each color is diluted with white and deepened with black to get various colors with different light and shade.

There should be a sample to control adjustment. First, prepare the sample. Find the base color and white, black matching proportion, and then prepare the required amount of paint according to the proportion. Generally, the paint color appears lighter when it is wet and darker when it is dry. This should be paid attention to when matching colors. When adjusting the color that compared with sample, it should be in the place where natural light is more sufficient, lest cause color difference.

New Words and Expressions

brand [brænd] *n.* 品牌；类型
blurred [blɜːd] *adj.* 模糊不清的
gelling [ˈdʒelɪŋ] *n.& v.* 胶凝；凝胶化
malpractice [ˌmælˈpræktɪs] *n.* 弊病
stir [stɜː(r)] *v.* 搅动；搅拌
settle [ˈsetl] *v.* 沉降
dispersion [dɪˈspɜːʃn] *n.* 分散；驱散
batten [ˈbætn] *n.* 木条
pneumatic [njuːˈmætɪk] *adj.* 气动的
viscosity [vɪˈskɒsəti] *n.* 黏度；黏性
harden [ˈhɑːdn] *v.* 硬化；变硬

scrap [skræp] *n.* 废料；*v.* 废弃
fluidity [fluˈɪdəti] *n.* 流动性
proportion [prəˈpɔːʃn] *n.* 比例
dilute [daɪˈluːt] *v.* 稀释
skinning [skɪnɪŋ] *n.* 结皮
filter [ˈfɪltə(r)] *v.* 过滤；滤网
impurity [ɪmˈpjʊərəti] *n.* 杂质；污点
sample [ˈsɑːmpl] *n.* 样板；样品
sufficient [səˈfɪʃnt] *adj.* 充足的；足够的

Notes

1. Before opening the paint pots, it is necessary to check carefully and confirm whether the variety, brand, color, and factory date meet the specified requirements.
   Translation: 涂料开罐前，应仔细检查并确认涂料的品种、牌号、颜色、出厂日期等是否符合规定要求。
2. The coating that cured by chemical reaction generally is a two-component type (even a three-component type), and the binder and curing agent of two-component coatings are packaged respectively, and mixed when in use.
   Translation: 一般靠化学反应而固化的涂料均为双组分型（甚至三组分型），双组分涂料的基料与固化剂分别包装，在使用时才混合。

Expanding Reading

Because the specific process procedure of ship construction is different from the production of general industrial products, it determines that the ship painting engineering should also be adapted to the shipbuilding process procedure. And it is different from that of general industrial products. Generally, during the whole process of shipbuilding, painting is divided into the following process stages:

Steel pretreatment and painting shop primer;

Block painting;

Slipway painting;

Wharf painting;

In-dock painting;

Outfitting parts painting.

The working characteristics and working essentials of the primer process stage of steel pretreatment and painting workshop have been introduced in Project 3, which will not be repeated here.

Block painting is the most important and basic part of ship painting. In addition to the special parts of special ships (such as the cargo tank of refined oil ships), all parts of the hull should be coated in part or all in the block stage.

Slipway painting refers to the painting operation in the process after closing on the platform until the ship is launched. The main work contents of painting at this stage is the repair of large joints between blocks and the repair of coating damage parts caused by mechanical reasons or welding or fire reasons after block painting. Meanwhile, the painting of parts that must be painted to a certain stage or all before the launch.

Wharf painting is the painting of the ship docked at the dock for outfitting operation. This stage shall be complete in all parts except for painting which must be performed in the dock.

The painting in the dock is mainly for the integrity painting of the lower area of the hull waterline, and also for some painting work too late in the outfitting stage of the dock.

For outfitting on board, whether large outfitting or small outfitting, most coats will be primer before installation, and the finish coat will be painted after installation.

Exercises

Ⅰ. Answer the following questions according to the passage.

1. What are the preparation work before painting?

2. A brief description of the color adjustment notes.

Ⅱ. Practice these new words.

1. English to Chinese.
    gelling _____   dispersion _____
    viscosity _____   fluidity _____
2. Chinese to English.
    搅拌 _____   固化 _____
    过滤 _____   结皮 _____

Ⅲ. Translation.

1. Translate the following sentences into Chinese.

    (1) When stirring, turn up the pigment deposited at the bottom of the pot with smooth and

clean batten, and then stir with an electric or pneumatic mixer until the viscosity of the coatings is uniform.

(2) The viscosity of the coatings has been adjusted in the factory. It can be used after being opened and stirred evenly.

2. Translate the short passage.
There should be a sample to control adjustment. First, prepare the sample. Find the base color and white, black matching proportion, and then prepare the required amount of paint according to the proportion. Generally, the paint color appears lighter when it is wet and darker when it is dry. This should be paid attention to when matching colors. When adjusting the color that compared with sample, it should be in the place where natural light is more sufficient, lest cause color difference.

Task 4.2 Brush Painting and Roller Painting

Text Reading

In the ship painting operation, brush painting, roller painting, compressed air spraying and high-pressure airless spraying are mainly adopted. Among these methods, the high-pressure airless spraying is the most widely used in the ship painting operation with its particular high painting efficiency. The characteristics and working essentials of each painting method will be described one by one below.

1. Brush painting

Brush painting is one of the simplest manual painting method. Brush painting tools are simple, and easy to operate and flexible. Due to its strong adaptability, the application is more common. But brush painting is time-consuming also more laborious because of its low work efficiency. So it is not often used in large area construction.

The advantages of brush painting are that the spraying tools are difficult to reach and ensure the thick film, such as various grooves, drainage holes, edges of ventilation vent, the sides of the steel and narrow areas. Before the large area of spraying, the above parts should be brushed painting with 1 to 2 layers.

There is a strong penetration force for brush coating, and it can make the paint penetrate into the fine holes and cracks. When there is a small amount of moisture on the coated surface, the brush painting can squeeze out the water, so that the painting can better adhere to the surface.

Another advantage of brush painting is that the paint is less wasteful and polluted to the environment.

For coatings with fast dry and poor leveling, brushing is not appropriate, which is easy to leave obvious brush marks and affect the flatness and beauty of the coating.

During brush painting, dip the paint brush with an appropriate amount of paint, and allocate to be in quite be coated surface range. Extend horizontally first, then straighten out vertically. They can make coating distribution evenly. You can also expand vertically first and then straighten out horizontally. Step by step and do not miss.

After paint brush is used, if there are physical curing coatings, the paint can be discharged as far as possible, then let vertical suspension immerse in clear water. When it will be used next time, the water can be shaken off. If there are chemical curing coatings, the solvent should be washed and then vertically suspended in the same solvent. Be careful not to make the brush dry, bent or messy.

2. Roller painting

Roller painting is suitable for painting of large planes that are difficult to spray for some reasons (Figure 4.3).

Figure 4.3　Roller painting

Roller painting efficiency is lower than spraying, but higher than brushing.

Roller painting is also less wasteful and polluted to the environment.

Another outstanding advantage of roller painting is that it can be operated over a long distance, reducing a part of the trouble of building scaffolding.

Similar to the brush painting, the roller painting also has a good penetration force.

However, for complex structures and uneven surfaces, the roller painting mode is limited. So the roller painting is often used in the outer plate, deck and superstructure outside surface of the hull.

The maintenance method of the paint roller is the same as that of the paint brush.

New Words and Expressions

brush painting 刷涂
roller painting 辊涂
compressed air spraying 压缩空气喷涂
high-pressure airless spraying 高压无气喷涂
manual [ˈmænjuəl] adj. 用手的；手工的
narrow [ˈnærəʊ] adj. 狭窄的；有限的
squeeze [skwiːz] v. 挤压
environment [ɪnˈvaɪrənmənt] n. 环境
dip [dɪp] v. 蘸
suspension [səˈspenʃn] n. 悬浮

Notes

1. Brush painting is one of the simplest manual painting method. Brush painting tools are simple, and easy to operate and flexible. Due to its strong adaptability, the application is more common. But brush painting is time-consuming also more laborious because of its low work efficiency. So it is not often used in large area construction.
   Translation: 刷涂是一种最简单的手工涂装方式。刷涂工具简单、操作方便、灵活、适应性强、应用较为普遍。但刷涂费时也较费力，工作效率低，因而在大面积施工中不常使用。

2. Roller painting is suitable for painting of large planes that are difficult to spray for some reasons.
   Translation: 辊涂适合于因某些原因而难以喷涂的大平面的涂装。

Expanding Reading

Features of Brush Painting and Roller Painting

Painting method	Principle	Feature	Use	Task environment	Pollution nuisance
Brush painting	Paint the brush with a painting	No special equipment, the use is very simple, but the painting efficiency is low, and easy to generate brush and marks and spots	It is widely used and available in all fields	Will not form coating dust, in good working environment	No coating dust, no pollution
Roller painting	Roll and painting on a cylinder made of sponge and other materials	Compared with the brush painting, although the efficiency is high, the paint film surface is not smooth, or curved surface can not be coated	Suitable for the outer wall and inner wall of the building	As same as a brush to painting, and the working environment is good	There is almost no public nuisance problems

Exercises

I. Answer the following questions according to the passage.

1. Can you describe the differences between brush painting and roller painting?

2. Can you briefly introduce the use of brush painting and roller painting?

II. Practice these new words.

1. English to Chinese.
   brush painting _____   roller painting _____
   compressed air spraying _____   high pressure airless spraying _____

2. Chinese to English.
   手工的 _____   蘸 _____
   狭窄的 _____   悬浮 _____

III. Translation.

1. Translate the following sentences into Chinese.

   (1) The advantages of brush painting are that the spraying tools are difficult to reach and ensure the thick film, such as various grooves, drainage holes, edges of ventilation vent, the sides of the steel and narrow areas. Before the large area of spraying, the above parts should be brushed painting with 1 to 2 layers.

   (2) Roller painting efficiency is lower than spraying, but higher than brushing. Roller painting is also less wasteful and polluted to the environment. Another outstanding advantage of roller painting is that it can be operated over a long distance, reducing a part of the trouble of building scaffolding.

2. Translate the short passage.

   After paint brush is used, if there are physical curing coatings, the paint can be discharged as far as possible, then let vertical suspension immerse in clear water. When it will be used next time, the water can be shaken off. If there are chemical curing coatings, the solvent should be washed and then vertically suspended in the same solvent. Be careful not to make the brush dry, bent or messy.

Task 4.3 Compressed Air Spraying

Text Reading

Compressed air spraying is to use the compressed air to attract (or press) the coating from the pot to the spray gun. Under 0.2 to 0.5 MPa, the coating is mixed with the air at the nozzle and sprayed on the coated surface to get an evenly distributed coating.

Compressed air spraying is much more efficient than brush painting and roller painting, and it is also easy to obtain a more uniform coating.

The coatings used for compressed air spraying are thinner than the brush painting and roller painting, so some thinner needs to be added into the general coating. In spraying, a part of the paint will drift into the air and from the coated surface rebound to the air. Therefore, the waste of paint is larger. Especially when there is a strong wind, the paint is blown out to cause more waste. The pollution impacting on the environment is obvious, and it does not brush the permeability, so the coating adhesion is not as good as brush painting and roller painting.

During spraying, the pressure of compressed air used, diameter and angle of the nozzle, the spraying distance and the amount of diluent can be adjusted according to the requirements of the coating product specification. The speed of the spray gun can be operated according to the film thickness requirements and the film formation situation.

The compressed air used must be filtered to remove moisture and oil to avoid affecting the coating quality.

The spray guns, containers and pipe tubes of the coating shall be washed with solvent immediately after spraying for later use.

The main tool of air spraying is the air spraying gun (commonly known as the spout gun). There are two kinds of common air spray guns.

1. Small pot spray gun

It is also known as the nozzle spray gun. It is composed of the gun body and storage tank, and the gun body is attached with the opening and closing of the compressed air valve trigger, air nozzle, and paint nozzle. Air nozzle and paint nozzle are perpendicular to each other. The storage tank attached to the gun body is equipped with return air holes, so that the paint can be ejected smoothly.

When the paint is added to the storage tank, you should contact the compressed air hose, and press the trigger, then the air flowing from the air nozzle will generate negative pressure in the suction pipe. The paint will be blown into fog with the compressed air flowing to the coated surface.

2. Large-sized spray gun

It is also known as flat nozzle spray gun or fan spray can gun. It is also composed of the gun body and the storage tank, but the gun body structure is smaller and more complex. After the compressed air enters the gun body, there are two ways passing through the air guide hole into the nozzle with the inner and outer layers of air ring wall. Its coating atomization principle is to use the special structure of the inner air ring wall and guide hole, so that the compressed air flows through the air hole and air ring wall, forming annular air flowing from the nozzle spray coating into small particles, and into a conical paint flow which is flown by the two air flows on both sides of the nozzle blown into a fan. The amplitude of the fan can be adjusted by the amplitude adjustment knob at the rear end of the gun body, and the flow of the paint can be controlled by the flow adjustment knob. Its storage capacity is large, therefore, it can be suitable for a larger area of painting.

New Words and Expressions

pot [pɒt] *n.* 罐；容器
nozzle [ˈnɒzl] *n.* 喷嘴
rebound [ˈriːbaʊnd] *vi.& vt.* 回弹
valve [vælv] *n.* 阀门
trigger [ˈtrɪɡə(r)] *n.* 扳机
perpendicular [ˌpɜːpənˈdɪkjələ(r)] *adj.* 垂直的
hose [həʊz] *n.* 软管
annular [ˈænjələ(r)] *adj.* 环形的
amplitude [ˈæmplɪtjuːd] *n.* 幅度
knob [nɒb] *n.* 球形把手；旋钮

Notes

1. Compressed air spraying is to use the compressed air to attract (or press) the coating from the pot to the spray gun.
   Translation: 压缩空气喷涂是利用压缩空气将涂料从壶形容器中吸引（或压迫）至喷枪。
2. During spraying, the pressure of compressed air used, diameter and angle of the nozzle, the spraying distance and the amount of diluent can be adjusted according to the requirements of the coating product specification. The speed of the spray gun can be operated according to the film thickness requirements and the film formation situation.
   Translation: 喷涂时，所用压缩空气的压力、喷嘴的口径和角度、喷涂距离和稀释剂添加量等，可根据涂料产品说明书的要求来调整。喷枪运行的速度可按膜厚要求与成膜情况而定。

Expanding Reading

Common Faults and Troubleshooting Methods of Compressed Air Spraying

Fault Phenomenon	Possible Cause	Exclusion Method
Spray gun leaks	The ball and the vent are not sealed; the nozzle is not tightened	Give the ball around the pores with sand wax. Tighten the nozzle and note the interface
Trigger fails	Dirt blockage	Remove the control valve, remove dirt, and seal the air valve with sand wax
Poor spray	Coating viscosity is too large. The paint tank is out of paint. Compressed air pressure is too low. Excessive paint outputs. There is moisture in the compressed air	Add the appropriate dilutions. Fill paint. Adjust and maintain the pressure. Reduce pressure if there is excessive pressure. Remove the water in the compressed air gas package
Gun body leakage	Backend packing of needle valve fails. The pressing nut is too loose. The needle valve is worn	Replace the filler. Tighten the nut. Grind in sand wax to tight
Pressure vessel leaks or leaks oil	The rolling head is too loose or the gasket is worn	Tighten the rolling head or replace the gasket
Discharge unilateral	The nozzle is blocked	Remove the nozzle, soak in a solvent, and blow through

Exercises

I. Answer the following questions according to the passage.

1. Please briefly describe the characteristics of compressed air spraying.

2. What are the main tools of compressed air spraying?

II. Practice these new words.

1. English to Chinese.
   nozzle _____   rebound _____
   valve _____   hose _____

2. Chinese to English.
   罐 _____   回弹 _____
   扳机 _____   幅度 _____

III. Translation.

1. Translate the following sentences into Chinese.

   (1) Compressed air spraying is much more efficient than brush painting and roller painting, and it is also easy to obtain a more uniform coating.

   (2) The compressed air used must be filtered to remove moisture and oil to avoid affecting the coating quality.

2. Translate the short passage.

   The coatings used for compressed air spraying are thinner than the brush painting and roller painting, so some thinner needs to be added into the general coating. In spraying, a part of the paint will drift into the air and from the coated surface rebound to the air. Therefore, the waste of paint is larger. Especially when there is a strong wind, the paint is blown out to cause more waste. The pollution impacting on the environment is obvious, and it does not brush the permeability, so the coating adhesion is not as good as brush painting and roller painting.

Task 4.4 High-Pressure Airless Spraying

Text Reading

High-pressure airless spraying usually uses compressed air as a power to drive the high pressure pump, inhale and pressurize the paint from 10 to 25 MPa through the high pressure hose and spray gun, and finally through the olive hole shaped nozzle spray. When the coating leaves the nozzle, it melts into very fine particles and sprays onto the coated surface, forming a uniform coated film.

Because the coating is pressurized to the high pressure by the high pressure pump, and the coating itself is not mixed with the compressed air, it is completely different from the compressed air spraying with the compressed air atomization coating. Therefore, it is called the high-pressure airless spraying.

Due to the high pressure of the coating, the coating can penetrate into the fine hole after shooting into the coated surface. So the coating adhesion is good. A smooth and dense coating can be earned to be good atomization.

The biggest advantage of high-pressure airless spraying is high efficiency which is dozens of times higher than brush painting or roller painting. For the ships that need a large area of painting, the painting efficiency is greatly improved, which can shorten a certain shipbuilding cycle. Therefore, modern shipbuilding is inseparable from high-pressure airless spraying.

The disadvantage of high-pressure airless spraying is that the escape loss of the coating is larger. Especially in the case of strong wind, the coating blowing loss gets larger. If the spraying surface shape is complex, or the width of objects is small, the loss is also larger. Compared with brush painting, high-pressure airless spraying usually requires from 20% to 30% more coatings.

In addition, because the coating at the nozzle outlet pressure is very high and the speed of the shooting is very high, it is easy to puncture the skin and cause damage. So it should be paid special attention.

During using high-pressure airless spraying, the following issues should be emphasized:

1. The pressure of spray and flow rate

For a certain type of airless spraying machine, when the used coating viscosity is constant, the input compressed air pressure meets a certain amount, there is a certain relationship between the spraying pressure and flow. Generally, when the flow rate increases, the spray pressure decreases. However, for the same airless spraying machine, if the coating viscosity is constant, when the pressure of the input compressed air increases, the spraying pressure and flow will increase accordingly.

2. Coating viscosity and spraying pressure

Different coatings have different viscosity. The higher viscosity of the coating, the greater spraying pressure is required. The viscosity of various coatings and the required spraying pressure can be found in the coating product specification.

It should be noted that the pressure of the high pressure pump output is different from the pressure of the nozzle outlet. The latter one is always lower than the former one, because the coating is rubbed in the high pressure hose, and the pressure will be dropped. The pressure drop size is related to the diameter and length of the hose. The smaller pipe diameter and the longer pipeline, the greater pressure drop. In addition, the pressure drop is also related to the flow rate, the greater flow rate, the greater pressure drop.

3. Nozzle selection

Before the high-pressure airless spraying, a certain aperture and a certain shaped nozzle should be selected according to the coating and spraying object used.

The aperture of the nozzle determines the flow rate, and the shape of the nozzle determines the size of the spray pattern.

The nozzle holes are mostly oval in shape. Some manufacturers provide the aperture size. The size is the diameter of a circle which equals to the oval. Some manufacturers do not provide the aperture size, but provide the flow data under certain conditions. For the coatings with higher viscosity, the nozzle with larger aperture or larger flow rate should be selected.

4. Spray the essentials

The spraying distance is typically between 30 cm and 50 cm. If the distance is too long, it will cause the paint film surface rough, and the loss of coating is also large. For some dry coatings, it will produce dry spray phenomenon. When the fog particle of the coating gets to the coated surface, it has become a dry powder state. If the distance is too short, the operation will be difficult and it is easy to thick paint film and cause flow sagging and wrinkle skin and other ills.

The spray gun should keep vertical to the coated surface. When the spray gun moves up and down, attention should be paid to keep a equal distance with the coated surface, avoiding arc or curve movement, in order to keep a the film thickness uniform.

The speed of the spray gun should be determined according to the requirements of film thickness. If the film thickness requirement is low, the moving speed will be faster, and if the film thickness requirement is high, the moving speed will be slower. Even back and forth or cross overlapping spraying should be carried out, sagging is not allowed. When spraying, the wet film thickness should be controlled by the wet film thickness meter, so as to meet the film thickness requirements. Not too thick or it will cause waste. For more complex corners and surface shape parts, intermittent rapid spraying is appropriate.

Generally, the spraying work should follow the principle of "up before down, and difficult before easy".

New Words and Expressions

pump [pʌmp] *n*. 泵
atomization [ˌætəmaɪˈzeɪʃən] *n*. 雾化
escape [ɪˈskeɪp] *v*. 喷逸；逸出
rub [rʌb] *v*. 摩擦
aperture [ˈæpətʃə(r)] *n*. 孔径
hang [hæŋ] *v*. 流挂
wrinkle [ˈrɪŋkl] *n*. 皱纹
overlap [ˌəʊvəˈlæp] *n*. 重叠
intermittent [ˌɪntəˈmɪtənt] *adj*. 间歇的；断断续续的
rapid [ˈræpɪd] *adj*. 瞬间的；快速的

Notes

1. High-pressure airless spraying usually uses compressed air as a power to drive the high pressure pump, inhale and pressurize the paint from 10 to 25 MPa through the high pressure hose and spray gun, and finally through the olive hole shaped nozzle spray. When the coating leaves the nozzle, it melts into very fine particles and sprays onto the coated surface, forming a uniform coated film.

   Translation: 高压无气喷涂通常是利用压缩空气作为动力驱动高压泵，将涂料吸入并加压至10~25 MPa，通过高压软管和喷枪，最后经呈橄榄形孔的喷嘴喷出。当涂料离开喷嘴时，雾化成很细的颗粒，喷射到被涂表面，形成均匀的涂膜。

2. The biggest advantage of high-pressure airless spraying is high efficiency which is dozens of times higher than brush painting or roller painting. For the ships that need a large area of painting, the painting efficiency is greatly improved, which can shorten a certain shipbuilding cycle. Therefore, modern shipbuilding is inseparable from high-pressure airless spraying.

   Translation: 高压无气喷涂的最大优点是效率高，比刷涂或辊涂高几十倍。对于需要大面积涂装的船舶来说，涂装效率的大大提高，可缩短一定的造船周期。因此，现代化的造船离不开高压无气喷涂。

Expanding Reading

Characteristics of Compressed Air Spraying and High-Pressure Airless Spraying

Paining method	Principle	Feature	Use	Task environment	Pollution nuisance
Compressed air spraying	The compressed air is used to aerosolize the coating and spray it on the surface to form the paint film	Equipment is cheap, range of use is wide and operation is simple. But it has the disadvantages of fog flying and coating loss	Used widely and can be used in all fields	The coating caused by compressed air and solvent evaporation, needs exhaust device, to pay special attention to sparks	Coating dust is particularly much, so we must consider the surrounding environment. In addition, the chimney at least should be higher than the roof, and the waste water excluded to the sewer. We must be careful not to make the solvent vapor accumulated
High-Pressure airless spraying	Add about 10~25 MPa hydraulic pump to spray atomization from the fine hole to form a paint film like compressed air spraying	Coating does not contain air. Fog scattered is less with high efficiency, but in the paint film surface and operation, it is not as good as compressed air spraying	Suitable for a large number of painting, can be used in all fields, but small pieces and focus on decoration operations are difficult to use	Although it is less than compressed air spraying, there is flying dispersion, and solvent volatilization still exist. So we still need the same device when compressed air spraying	Almost the same as air spraying. During outdoor spraying, windy dust will attach to the car, and dry things above. So pay more attention when carrying out outdoor operation under the wind

Exercises

I. Answer the following questions according to the passage.

1. Please briefly describe the characteristics of high-pressure airless spraying.

2. What problems should be paid attention to when applying high-pressure airless spraying?

II. Practice these new words.

1. English to Chinese.

   atomization _____ escape _____

   overlap _____ rapid _____

2. Chinese to English.

   泵 _____ 雾化 _____

   流挂 _____ 重叠 _____

III. Translation.

1. Translate the following sentences into Chinese.

   (1) Due to the high pressure of the coating, the coating can penetrate into the fine hole after shooting into the coated surface. So the coating adhesion is good. A smooth and dense coating can be earned to be good atomizations.

   (2) The disadvantage of high-pressure airless spraying is that the escape loss of the coating is larger. Especially in the case of strong wind, the coating blowing loss gets larger. If the spraying surface shape is complex, or the width of objects is small, the loss is also larger. Compared with brush painting, high-pressure airless spraying usually requires from 20% to 30% more coatings.

2. Translate the short passage.

   The aperture of the nozzle determines the flow rate, and the shape of the nozzle determines the size of the spray pattern.

   The nozzle holes are mostly oval in shape. Some manufacturers provide the aperture size. The size is the diameter of a circle which equals to the oval. Some manufacturers do not provide the aperture size, but provide the flow data under certain conditions. For the coatings with higher viscosity, the nozzle with larger aperture or larger flow rate should be selected.

Project 5 Coating Quality Inspection and Control

Background

The quality of the coating is related to many factors, such as the corrosion degree of the substrate surface, the types of the substrate surface pollutants, rust removal methods, rust removal grades, coating types, composition, coating methods, coating environment, etc. However, no matter how much attention is paid to each link, it is impossible to avoid coating problems, resulting in defects or damage to the final coating surface. If these defects are not handled correctly, it will affect the protective effect of the coating, and then affect the service life of the substrate, and even lead to serious accidents.

In order to maximize the protection of the ship, extend the service life of the coating, and implement the most effective anti-corrosion measures for the substrate, we should first understand the types of coating defects, and then give solutions according to the state of the defect, and take effective repair measures. Since the corrosion and protection methods of steel structure substrate have been introduced before, this project mainly introduces the common defects of coating and repair measures through state, color and scope, etc. for determining the causes of the formation of defects, and then gives solutions and suggestions. In addition, several common detection methods and steps of coating performance are introduced to help to understand and master the use and operation of coating quality inspection equipment.

Learning Target

Knowledge Target

1. Understand the words and sentence patterns related to coating defects.
2. Master the vocabulary and usage related to common coating defects and repair measures.
3. Master the vocabulary and usage of the field detection methods and steps of coating defects.

Ability Target

1. Skillfully use the coating quality inspection and other related words.
2. Be able to accurately locate and express the coating defects in English.

Task Definition

Task 5.1 Introduction of Common Defects on Coating Surface 1

Task 5.2　Introduction of Common Defects on Coating Surface 2
Task 5.3　Field Detection Methods and Steps of Coating Performance

Task 5.1　Introduction of Common Defects on Coating Surface 1

Text Reading

　　Ship painting, whether new shipbuilding or ship repair, is affected by many factors. Apart from the previously introduced substrate defects, due to the coating itself and painting methods, painting environment problems are not a minority. This task mainly introduces the common defects on the coating surface, the defect formation mechanism, repair suggestions and methods, etc.

　　1. Sagging (Figure 5.1)

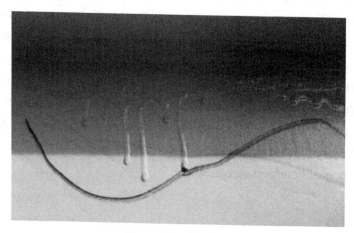

Figure 5.1　Sagging

　　The defect formation mechanism is as follows:

　　(1) The nozzle pressure of the spray gun is too large, or the spray gun is too near the coated surface, causing the unit area of the substrate surface covering excessive coating. If there is a too thick coating, it will not bear the action of gravity and slide down, forming sagging.

　　(2) The nozzle is worn and damaged, resulting in the increase of coating output per unit time (larger aperture, smaller spray range), the paint film covered by the substrate surface is too thick and slides down, forming sagging.

　　(3) During site construction, excessive addition of diluent, the substrate surface temperature or the temperature of the coating itself is too high, resulting in a reduced viscosity of the coating, and sliding off after spraying, forming sagging.

　　(4) Coating failure, when the coating finishes the shelf life or effective mixing time, it may reduce the viscosity of the coating, after spraying the coating slide.

Recommendations and patching methods

Generally, the slight sagging only affects the appearance of the coating and does not affect the other properties of the coating. Whether to repair it can be based on the actual situations. More serious sagging will lead to local paint film to be too thick. If the single number of lays paint film is too thick, it may affect the drying of local paint film and insufficient dry paint film may affect the final waterproof, chemical prevention and other properties, etc. When the paint film slips down, the local coating slides down and causes the paint film thin, and even exposes to the bottom surface. The lower part accumulates due to coating sliding, which must be repaired. The most common repair method is sandpaper grinding (thicker flow hanging paint film can be used to carefully remove the protruding paint film), and then cover the thinner coating with one number of lay which can improve the surface beauty of the repair part.

2. Pinhole (Figure 5.2)

Figure 5.2 Pinhole

Thicker coating is covered on the porous substrate surface (such as metal thermal spraying, inorganic zinc silicate coating, and concrete). After the air overflows in the gap, the coating cannot close the air overflow channel and form a discontinuous coating. Especially under the forced drying conditions of paint film, it is more likely to appear.

Recommendations and patching methods

Because the pinhole generally penetrates through the whole coating, it is necessary to completely remove the pinhole defects through mechanical treatment methods, such as abrasive grinding or power tool grinding methods to completely remove the pinhole defects, and then respray coatings. If only the coating surface is covered and repaired, the effective protective performance of the coating cannot be obtained.

3. Bubble (Figure 5.3)

The thicker coatings cover the porous substrate surface, and the air in the pores overflows but does not permeate the paint film, just tops the surface coating, forming an arc bulge, especially under forced drying conditions.

Figure 5.3   Bubbles

Recommendations and patching methods

Bubble and pinhole essence are the same defects. Both of them are in the paint film formed fine pores, so the bubble defect repair is the same as pinhole, which should be through mechanical treatment methods, such as abrasive grinding or power tool grinding methods to completely remove bubble defects, and then spray coating. Only polishing the coating surface can not obtain the effective protective performance of the coating.

4. Wrinkle (Figure 5.4)

Figure 5.4   Wrinkle

Coatings (alcoholic acid, and modified alcoholic acid coating varieties) single spraying thickness is too large. Especially in low temperature conditions, film curing reaction is slow. In this case, the curing reaction on the surface of coating, and the solvent inside the coating is still volatilizing. The volatilized solvent will shrink the partially cured coating made on the surface, and eventually cause the coating wrinkle and form a wrinkled paint film.

Recommendations and patching methods

Slight wrinkle appearance can not be repaired, because slight paint film wrinkle will not affect the protective performance of the coating. If it is a serious wrinkle problem, we need to use mechanical treatment method (abrasive grinding or power grinding methods) to remove the surface wrinkle, and then spray once as a cover. It is the best to be in the construction of alcoholic acid coating. Do not spray too thick, which could avoid paint film wrinkle.

New Words and Expressions

defect [ˈdiːfekt] *n.* 缺陷
sag [sæg] *vi.* 下垂；*vt.* 使……下垂
gravity [ˈɡrævəti] *n.* 重力
slide [slaɪd] *v.* 滑动
wear [weə(r)] *v.* 磨损
aperture [ˈæpətʃə(r)] *n.* 孔；洞
failure [ˈfeɪljə(r)] *n.* 失效；失败
recommendation [ˌrekəmenˈdeɪʃn] *n.* 推荐；建议
waterproof [ˈwɔːtəpruːf] *adj.* 防水的
accumulation [əˌkjuːmjəˈleɪʃn] *n.* 堆积；积累
sandpaper [ˈsændpeɪpə(r)] *n.* 砂纸
eradicate [ɪˈrædɪkeɪt] *v.* 铲除；根除
protrud [prəˈtruːd] *vt.* 突出；伸出
pinhole [ˈpɪnhəʊl] *n.* 针孔
bubble [ˈbʌbl] *n.* 气泡
overflow [ˌəʊvəˈfləʊ] *vt.&vi.* 溢出
wrinkle [ˈrɪŋkl] *n.* 皱纹；*v.* 使……起皱
alcoholic acid 醇酸
volatile [ˈvɒlətaɪl] *adj.* 易挥发的
shrinkage [ˈʃrɪŋkɪdʒ] *n.* 收缩

Notes

1. The nozzle pressure of the spray gun is too large, or the spray gun is too near the coated surface, causing the unit area of the substrate surface covering excessive coating. If there is a too thick coating, it will not bear the action of gravity and slide down, forming sagging.
   Translation: 涂料喷枪的喷嘴压力过大，或者喷枪太靠近被涂表面，造成底材表面的单位面积覆盖过量涂料。过厚涂料无法承受重力作用而向下滑落，形成流挂。

2. The thicker coatings cover the porous substrate surface, and the air in the pores overflows but does not permeate the paint film, just tops the surface coating, forming an arc bulge, especially under forced drying conditions.
   Translation: 较厚涂层覆盖在多孔的底材表面，孔隙内的空气溢出但未渗透过漆膜，只是顶起表层涂料，形成圆弧状的凸起，尤其在强制烘干条件下更容易出现这种情况。

Expanding Reading

There are also some defects in the regular construction. The following will be listed one by one.

1. Over-spray and paint mist (Figure 5.5)

Figure 5.5　Paint mist

In the process of coating spraying, excessive ventilation or excessive outdoor wind will cause paint dust and paint mist pollution to the surrounding surface that has completed painting construction, forming unsightly paint mist pollution. Usually, the spray paint requires the spray gun to equal distance horizontal move with the coated surface, and the wrong spraying technique or operation (such as arc spraying gesture) may cause the spray paint mist pollution to the surrounding surface where the coating construction has been completed.

Recommendations and patching methods

Reasonable ventilation conditions must be ensured before spraying, and environmental wind factors should be taken into account during outdoor construction. Meanwhile, during the spraying process, it is necessary to consider appropriately covering the structural surface around the spraying.

2. Fish eye

Fish eye is a special case of pit defects, which is mainly due to silicone oil pollution. The incompatibility between silicone oil and coating makes the coating can not moisten the coating contaminated by silicon oil. Paint film is pushed away to form a pit, and there is a black dot in the center of the hole (similar to the fish eye).

Recommendations and patching methods

The main sources of silicone oil pollutants are: the coating itself (may be contaminated in the production process or coating formulation problems) or the coating in the construction process (such as the pollution of construction equipment), or the spraying surface by the external environment pollution. The exact causes can be judged by site analysis. Fish eye defects cannot be removed and repaired simply by mechanical treatment, because silicone oil as a contaminant can not be removed completely by mechanical method only. Disposable abrasive should be used to remove the coating with fish eye defects (treated abrasive is not recommended continuing to use). The surface is then thoroughly cleaned with a diluent with the aim of removing silicone oil contaminants invisible to the naked eye. Finally, repaint it.

3. Shaft hole

There are oil or water vapor pollutants on the spraying surface (or in the coating), resulting in the subsequent wet film of the coating being pushed away because it cannot wet the polluted parts, forming a discontinuous volcanic pit.

Recommendations and patching methods

Conventional pits can be removed by mechanical treatment methods (such as abrasive blasting or mechanical grinding methods). The pits caused by serious oil pollution must be removed by chemical treatment. Simple mechanical treatment can not meet the repair expectations.

Exercises

Ⅰ. Answer the following questions according to the passage.

1. What are the fix methods for sagging?

2. A brief description of the repair methods of bubbles.

Ⅱ. Practice these new words.

1. English to Chinese.

   sag _____   pinhole _____
   bubble _____   wrinkle _____

2. Chinese to English.

   缺陷 _____   下垂 _____
   失效 _____   铲除 _____

Ⅲ. Translation.

1. Translate the following sentences into Chinese.

   (1) Thicker coating is covered on the porous substrate surface (such as metal thermal

spraying, inorganic zinc silicate coating, and concrete). After the air overflows in the gap, the coating cannot close the air overflow channel and form a discontinuous coating. Especially under the forced drying conditions of paint film, it is more likely to appear.

(2) Coatings (alcoholic acid, and modified alcoholic acid coating varieties) single spraying thickness is too large. Especially in low temperature conditions, film curing reaction is slow. In this case, the curing reaction on the surface of coating, and the solvent inside the coating is still volatilizing.

2. Translate the short passage.

Generally, the slight sagging only affects the appearance of the coating and does not affect the other properties of the coating. Whether to repair it can be based on the actual situations. More serious sagging will lead to local paint film to be too thick. If the single number of lays paint film is too thick, it may affect the drying of local paint film and insufficient dry paint film may affect the final waterproof, chemical prevention and other properties, etc. When the paint film slips down, the local coating slides down and causes the paint film thin, and even exposed to the bottom surface. The lower part accumulates due to coating sliding, which must be repaired. The most common repair method is sandpaper grinding (thicker flow hanging paint film can be used to carefully remove the protruding paint film), and then cover the thinner coating with one number of lay which can improve the surface beauty of the repair part.

Task 5.2 Introduction of Common Defects on Coating Surface 2

Text Reading

1. Lifting

Strong solvent or coating which contains strong solvent covering on the (alcoholic acid oxidation, curing, etc.) coating surface. Because the oxidation or curing reaction is not sufficient, (alcoholic acid oxidation curing) coating affected by strong solvent is bitten up or even spalling, and the phenomenon of folding appears.

Recommendations and patching methods

Remove damaged paint film with mechanical treatment and spray paint according to specification.

2. Orange skin (Figure 5.6)

Figure 5.6 Orange skin

Because the coating viscosity is too large, or the self-leveling ability is poor, the surface of the paint film after spraying is not well-formed, and the appearance is rough, which is similar to the state of the orange peel surface.

Recommendations and patching methods

Usually, orange peel defects only appear on the surface layer of the paint film and will not affect the protective performance of the coating. If it needs to be repaired, it can be grinded with fine sandpaper to remove the rough surface. Be sure to avoid leaving grinding marks during grinding, otherwise the repair result will be counterproductive.

3. Bleeding (Figure 5.7)

Light paint covers a coal tar coating or a coating surface containing tar component, and the bottom tar components will run into the surface of coating freely, making the light surface layer brown or black.

Figure 5.7 Bleeding

Recommendations and patching methods

Usually the bleeding phenomenon only makes the color of the surface paint film deep. It will not affect the other properties of the paint film. Even if it is repaired, the light surface will appear bleeding phenomenon. However, if the surface of the bleeding paint film is covered with the subsequent coating later, and the seeping tar component (somewhat sticky) exists on the surface of the paint film, it will cause the adhesion problem among the coatings without cleaning up. So the seeping coating surface must be treated before coating. Usually, heated fresh water washing can achieve the treatment effect, and other mechanical methods of treatment of seepage color surface are not thorough.

4. Cracking (Figure 5.8)

Figure 5.8 Cracking

The most fundamental cause of coating cracking is that the paint film cannot eliminate the internal stress. Main stresses include:

(1) The coating or substrate is partially heated and the thermal stress is transmitted to the paint film;

(2) Paint film is too thick (especially thermoset coatings). The paint film is weak enough to eliminate stress;

(3) Changes of hot and cold, dry and wet cycles in the external environment and external stress are transmitted to the inside of the paint film;

(4) Thermoset coatings covered over the thermoplastic coating surface. Bottom thermoplastic coatings are transmitted to the upper thermosetting coatings through the temperature change;

(5) The stress generated by the lack of rigid structure substrate is transmitted to the paint film;

(6) Due to prolonged exposure to corrosion conditions, the coating is aging, brittle and losing toughness.

Recommendations and patching methods

If the coating occurs crack phenomenon, it can be considered that the coating has lost the protective performance of the original design, and must be fully repaired. Abrasive blasting treatment is the preferred treatment method, completely removes the cracking coating and achieves a certain cleanliness and roughness, finally sprays the coating according to the requirements.

5. Spalling (Figure 5.9)

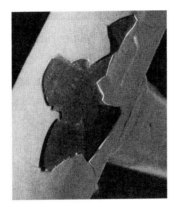

Figure 5.9　Spalling

The causes of coating spalling are: inadequate surface cleaning, such as pollutants before coating construction; too smooth surface or poor surface roughness; the coating/recoating interval of thermoset coatings exceeds the specified time limit; and poor compatibility among different coatings.

Recommendations and patching methods

Abrasive sand treatment is the preferred treatment method. The peeled coating shall be completely removed to achieve certain cleanliness and roughness, and then spray according to the requirements of each coating. If the surface pollutants are caused by surface spalling problem, it is necessary to conduct adhesion test to the coating around the peeling area in order to avoid the repair work.

New Words and Expressions

lifting [lɪftɪŋ] *n.* 咬底
bite up 咬起

spalling [ˈspɔːlɪŋ] *n.* 剥落
orange peel 橘皮
bleeding [ˈbliːdɪŋ] *n.* 渗色
seeping [siːpɪŋ] *v.* 渗出
eliminate [ɪˈlɪmɪneɪt] *v.* 排除；消除
thermoset [ˈθɜːməset] *n.* 热固性
thermoplastic [ˈθɜːməʊˈplæstɪk] *n.* 热塑性
embrittlement [emˈbrɪtlmənt] *n.* 脆化

Notes

1. Strong solvent or coating which contains strong solvent covering on the (alcoholic acid oxidation, curing, etc.) coating surface. Because the oxidation or curing reaction is not sufficient, (alcoholic acid oxidation curing) coating affected by strong solvent is bitten up or even spalling, and the phenomenon of folding appears.
   Translation: 强溶剂或含有强溶剂的涂料覆盖在（醇酸等氧化、固化类）涂层表面，因氧化固化反应不充分，（醇酸类氧化固化类）涂层受强溶剂影响被咬起甚至剥离，出现皱褶的现象。

2. Because the coating viscosity is too large, or the self-leveling ability is poor, the surface of the paint film after spraying is not well-formed, and the appearance is rough, which is similar to the state of the orange peel surface.
   Translation: 涂料由于黏度过大，或者自流平能力差，造成喷涂后漆膜表面成形不好，外观较粗糙，类似于橘子皮表面的状态。

Expanding Reading

1. Dry spray

Because the spraying surface temperature is too high (such as direct sunlight surface), or the spray distance is long (between spray gun and surface distance), the paint drying is too fast (add volatile dilution). The atomized paint drops of liquid coating are dried into fine solid dry power before reaching the sprayed surface, forming discontinuous coating power with no (or poor) adhesion and adhering to the substrate surface. This discontinuous coating has no protective performance, and will affect the attachment ability of subsequent coating.

Recommendations and patching methods

Mechanical treatment method can repair the dry spray coating, such as mechanical grinding. However, for inorganic zinc silicate coating, because the coating itself has poor recoating performance, the necessary abrasive blasting treatment to remove the dry coating layer is the only way to repair. The expected performance of the coating can be achieved only by removing all the coatings and then full spraying .

2. White film of paint (Figure 5.10)

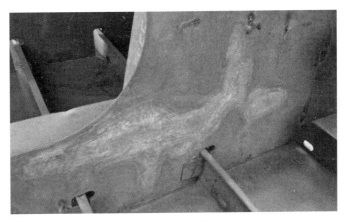

Figure 5.10   White film

Under the environmental conditions of high humidity and low temperature, the free hydrophilic curing agent in the amine-cured epoxy coating reacts with the condensed water vapor to produce oily reactants, which further reacts with carbon dioxide in the atmosphere and generates white carbaminate on the paint film surface, and whitening on the paint film surface.

Recommendations and patching methods

The amine-induced white coating can be washed by weakly alkaline hot water. If the coating is coated later, it must be cleaned with fresh water to avoid affecting the binding force among the coatings.

3. Powder

Because the organic coating resin (adhesive) is exposed to the corrosion environment (especially ultraviolet light) aging decomposition, the pigment wrapped in the resin is separated out, which makes the surface of the paint film powdery. Especially for the coatings with poor weather resistance, such as chlorinated rubber, and epoxy resin, are more likely to appear powder phenomenon.

Recommendations and patching methods

Powder phenomenon only appears on the coating surface. It does not affect the protective performance of the coating. In order to ensure the beauty of the surface, it can be washed with fresh water. If the coating surface needs to be covered other coatings, in order to ensure the adhesion among the coatings, the original coating must be properly treated. A larger treatment area can choose abrasive blasting, and a smaller area can choose manual sandpaper grinding. No matter which treatment method is used, it must be ensured that there is no power residue on the original coating surface before subsequent coating coverage.

Exercises

I. Answer the following questions according to the passage.

1. Can you describe the cause of lifting?

2. How to repair the coating problem caused by the bleeding?

II. Practice these new words.

1. English to Chinese.
   orange peel _____   eliminate _____
   bleeding _____   thermoset _____
2. Chinese to English.
   咬底 _____   剥落 _____
   热塑性 _____   脆化 _____

III. Translation.

1. Translate the following sentences into Chinese.

   (1) Light paint covers a coal tar coating or a coating surface containing tar component, and the bottom tar components will run into the surface of coating freely, making the light surface layer brown or black.

(2) Paint film is too thick (especially thermoset coatings). The paint film is weak enough to eliminate stress.

2. Translate the short passage.

The causes of coating spalling are: inadequate surface cleaning, such as pollutants before coating construction; too smooth surface or poor surface roughness; the coating/recoating interval of thermoset coatings exceeds the specified time limit; and poor compatibility among different coatings.

Task 5.3 Field Detection Methods and Steps of Coating Performance

Text Reading

The quality of the coating can be evaluated through some relevant testing methods. The most important evaluation indicators are wet film thickness, dry film thickness, adhesion, etc. This task mainly introduces several methods and steps of measuring film thickness and adhesion, which are also the most common and practical detection methods in shipyards at present.

1. Coating wet film thickness measurement

The comb tooth shape wet film card is mainly used to measure the wet film thickness of the coating. The measurement steps are as follows:

(1) Clean the comb wet film card with the appropriate method, and select the test range corresponding to the thickness of the wet film;

(2) When the coating is sprayed, immediately use the wet film card to measure the paint film thickness in order to avoid the drying of the coating and affect the accuracy of the reading;

(3) Read the wet film thickness reading and record the tooth mark thickness of the last one touching the paint and the first tooth mark thickness of not touching the paint.

2. Measurement of coating dry film thickness (Figure 5.11)

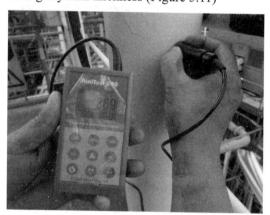

Figure 5.11 The measurement of dry film thickness

Elcometer 345# measuring instrument is mainly used to measure the dry film thickness of coating. The steps are as follows:

(1) Zero calibration of dry film meter: select a smooth steel plate with clean surface and no pollutants, then press the film thickness meter probe on the zero plate. Adjust the button to make that the reading is "0";

(2) Select the standard diaphragm or plate, according to the required measured coating thickness of the approximate value (and slightly above the measured thickness value). Place the

diaphragm or plate on the zero plate, and then press the probe on the above. Adjust the key to make the reading display with the standard diaphragm or plate thickness value;

(3) Repeat the above steps until the one-time read value is in agreement with the zero reading and the thickness value of the standard diaphragm or sample plate respectively, so that the zero calibration procedure of the film thickness table is completed;

(4) Measure the dry film thickness of the coating according to the specification requirements and record (the dry film meter is calibrated prior to each measurement or as determined by each party).

3. Coating adhesion test

There are two methods of common adhesion testing — grid test and pull method.

(1) Grid test

① Measure the test area with a calibrated dry film gauge and obtain an average thickness, and select the grid spacing according to the average value of the measured thickness (according to the standard);

② Cut 6 equidistant horizontal knives (up to the surface of the substrate), and then cut 6 knives perpendicular to the horizontal blade direction to form a 90° grid;

③ Gently brush the debris from the surface cutting, then touch the sticky tape on the grid, and completely compact the tape with the nails;

④ Remove the tape and attach it on the report paper (board), and rate the coating adhesion against the standard.

The specific adhesion level is different according to the cutter selected.

(2) Pull method

① Clean and brush slightly to test surfaces and column specimen (adhesion) surfaces;

② Apply daub glue evenly on the column specimen surface, then compact the surface of the coating test, stand for a period of time (according to the glue instructions) to let the glue be completely cured;

③ After the glue is completely cured, carefully cut the coating around the column specimen (until the substrate surface) with the ring cutting tool to avoid pulling the specimen during the cutting process;

④ Cover and lock the column specimen with the pull connector, pay attention to avoiding pulling the specimen and turning the button to align the red and black pointers in the dial to zero. Then turn the handle at a constant speed of about 1 MPa per second, and the specimen should be pulled out within 90 s;

⑤ After pulling out the column specimen, immediately read the reading indicated by the red pointer in the dial, and record the pull force value (in MPa);

⑥ Assess and record the coating damage and area (percentage) according to the appearance of the coating test surface and the column specimen surface and in the standard required format.

Finally, based on the ISO 4624, paint and varnish-adhesion test results according to the following table:

\multicolumn{2}{c}{Evaluation of the representation of the damage layer type}	
A	Internal destruction of substrate
A/B	Interdestruction between substrate and first degree coating
B	Inner destruction of first degree coating
B/C	Interbreakdown between the first and second degree coatings
n	Inner layer destruction of n degree
n/m	Interlayer damage between coatings at degrees n and m
-/Y	The final interlayer damage between the coating and the glue
Y	The damaged inside layer of the glue
Y/Z	Interdestruction between glue and specimen

New Words and Expressions

measurement [ˈmeʒəmənt] *n.* 测量

comb [kəʊm] *n.* 梳子

calibration [ˌkælɪˈbreɪʃn] *n.* 校准

probe [prəʊb] *n.* 探头

grid [grɪd] *n.* 格子；网格

equidistant [ˌiːkwɪˈdɪstənt] *adj.* 等距的

perpendicular [ˌpɜːpənˈdɪkjələ(r)] *adj.* 垂直的

sticky [ˈstɪki] *adj.* 黏性的

tape [teɪp] *n.* 胶带

nail [neɪl] *n.* 指甲

column [ˈkɒləm] *n.* 柱状物

specimen [ˈspesɪmən] *n.* 样品；范例

glue [gluː] *vt.* 胶合；*n.* 胶水

Notes

1. The comb tooth shape wet film card is mainly used to measure the wet film thickness of the coating.

   Translation: 涂层湿膜厚度测量，主要采用的是梳齿状湿膜卡。

2. Measure the dry film thickness of the coating according to the specification requirements and record (the dry film meter is calibrated prior to each measurement or as determined by each

party).

Translation: 根据规范要求测量漆膜涂层的干膜厚度,并记录(建议每次测量之前进行干膜测量仪的校准,或者事先经各方确定)。

Expanding Reading

Because the quality of the coating is greatly related to the substrate surface state, the quality inspection of the substrate surface treatment is also very important. Several typical substrate surface detection items are described below.

1. Detection of water-soluble salt on the substrate surface

Water-soluble salt on the substrate surface is mainly detected by Bresle stickers. The method is as follows:

(1) Clean the conductivity meter, syringe (needle) and container;

(2) Take a Bresle patch, strip off the protective pad on the back of the patch, and remove the pad from the test area. Then, press the patch to the surface test area (ensure as little residual air in the test area as possible)(Figure 5.12);

Figure 5.12 Sticking Bresle patch in test area

(3) Draw a certain amount of pure water (10 mL, 15 mL, 20 mL) with a needle into the container. Measure the original conductivity reading (unit: μs/cm) of the pure water, and record the reading (Figure 5.13);

Figure 5.13 Testing the electrical conductivity of pure water

(4) Extract about 3.5 mL of pure water from the container and insert the test area from the protective foam around the patch at about 30° angle (Figure 5.14);

Figure 5.14 Pouring pure water into the test area

(5) After injecting pure water into the test area and standing for 1 minute (only on the first injection), extract the solution from the test area, and then inject-extract the solution 9 times continuously. Finally, extract as much solution as possible into the syringe and inject it into the container;

(6) The conductivity of the test solution (in μs/cm) is measured with a conductivity meter, and the final actual conductivity is equal to the measured value minus the original conductivity of the pure water (Figure 5.15).

Figure 5.15 Solution conductivity test

Finally, the salt content of the substrate surface is calculated from the amount of pure water used on the test (Figure 5.16).

Figure 5.16 Calculation of salt content in test area

2. Roughness detection

(1) Select the test area and remove surface dust, then judge the rough shape (sharp or arc) of the sand surface by fingertip scraping or using 5-fold magnifying glass. If the material (angular sand or steel pill) of the abrasive is known, directly select the standard sample type (S-shot / arc,

G-angular sand / sharp);

(2) By scraping one fingertip at a time on the roughness level of the standard plate and the test area, the neatest surface roughness level is evaluated and the actual roughness level is determined. If it is necessary, you can also use a 5-fold magnifier to help to determine the roughness grade.

Finally, the roughness level is determined by surface roughness characteristics of the steel pre-treated by blasting before coating and related products according to ISO 8503.

Exercises

I. Answer the following questions according to the passage.

1. Please briefly describe the test method of paint film thickness.

2. What are the main test steps of coating adhesion?

II. Practice these new words.

1. English to Chinese.
   calibration _____   probe _____
   equidistant _____   grid _____

2. Chinese to English.
   测量 _____   黏性的 _____
   胶带 _____   柱状物 _____

III. Translation.

1. Translate the following sentences into Chinese.

(1) Measure the test area with a calibrated dry film gauge and obtain an average thickness, and select the grid spacing according to the average value of the measured thickness (according to the standard).

(2) After pulling out the column specimen, immediately read the reading indicated by the red pointer in the dial, and record the pull force value (in MPa).

2. Translate the short passage.

The quality of the coating can be evaluated through some relevant testing methods. The most important evaluation indicators are wet film thickness, dry film thickness, adhesion, etc. This task mainly introduces several methods and steps of measuring film thickness and adhesion, which are also the most common and practical detection methods in shipyards at present.

参考文献

[1] 陆伟东，连琏. 船舶与海洋工程专业英语 [M]. 上海：上海交通大学出版社，2009.
[2] 王建红. 船舶日常实用英语 [M]. 哈尔滨：哈尔滨工程大学出版社，2021.
[3] 彭公武. 造船专业英语 [M]. 哈尔滨：哈尔滨工程大学出版社，2006.
[4] 彭辉，王金鑫，司卫华. 船舶防腐与涂装 [M]. 哈尔滨：哈尔滨工程大学出版社，2014.
[5] 蒋一兵. 涂装检查参考手册 [M]. 北京：化学工业出版社，2014.